Earle Abbott Merrilol

Electric Lighting Specifications

For the Use of Engineers and Architects. Second Edition

Earle Abbott Merrilol

Electric Lighting Specifications
For the Use of Engineers and Architects. Second Edition

ISBN/EAN: 9783337249182

Printed in Europe, USA, Canada, Australia, Japan

Cover: Foto ©berggeist007 / pixelio.de

More available books at **www.hansebooks.com**

BY THE SAME AUTHOR

REFERENCE BOOK OF
TABLES AND FORMULAS
FOR
Electric Street Railway Engineers
Flexible Morocco.   Price, $1.00

THE W. J. JOHNSTON COMPANY

253 Broadway, New York

ELECTRIC LIGHTING SPECIFICATIONS

FOR THE USE OF

ENGINEERS AND ARCHITECTS

BY

E. A. MERRILL, A. M.,

*Author of "Reference Book of Tables and Formulas for
Electric Street Railway Engineers"*

Second Edition. Entirely Rewritten

NEW YORK
THE W. J. JOHNSTON COMPANY
253 BROADWAY
1896

COPYRIGHT, 1892 AND 1896,
BY THE W. J. JOHNSTON COMPANY.

CONTENTS.

	PAGE
Introduction	19
Working Outline	23

SPECIFICATIONS.

Warning	30

PREAMBLE.

Bids	31
Bond	32
Contractor	32
Commencement and Completion of Work .	32
Damages	34

GENERAL SPECIFICATION.

Duties of Contractors . .	34
Work, Labor, and Materials . . .	36
Additional, Omitted, or Changed Work .	36
Replacement of Defective Material . .	37

CONTENTS.

	PAGE
Patented Apparatus	37
Special Devices	37
Safeguards and Debris	37
Plans	38
Tests	38
Inspection	40
Insurance Rules	41
Acceptance	41
Terms of Payment	42

INSTALLATION OF DYNAMOS AND SWITCH-BOARDS. LOW POTENTIAL, DIRECT CURRENT SYSTEM, TWO-WIRE OR THREE-WIRE.

Dynamos	43
Foundations	46
Instruments	48
Cables to Switchboard	50
Starting Plant and Instruction	51
Renewal Parts	51
Summary	52
Switchboard	53
Switchboard Apparatus	54
Connections (concealed)	55

CONTENTS. vii

	PAGE
Connections (surface)	56
Circuits	57

INCANDESCENT SERIES SYSTEM, DYNAMOS MEDIUM OR HIGH POTENTIAL, (VARIABLE OR CONSTANT), CURRENT DIRECT OR ALTERNATING.

Dynamos	58
Foundations	61
Instruments	62
Cables to Switchboard	63
Starting Plant and Instruction	64
Renewal Parts	64
Summary	64
Switchboard	65
Switchboard Apparatus and Connections	65
Circuits	66

CONSTANT POTENTIAL, ALTERNATING CURRENT SYSTEM.

Dynamos	67
Foundations	69
Instruments	69
Converters	71

CONTENTS.

	PAGE
Cables to Switchboard	71
Starting Plant and Instruction	71
Renewal Parts	71
Summary	72
Switchboard	72
Switchboard Apparatus and Connections	72
Circuits	73

ALTERNATING CURRENT OR DIRECT CURRENT SYSTEM WITH THE PARALLEL SYSTEM OF DISTRIBUTION.

Dynamos	74
Foundations	76
Instruments	78
Converters	79
Cables to Switchboard	79
Starting Plant and Instruction	81
Renewal Parts	81
Summary	81
Switchboard	82
Switchboard Apparatus and Connections	82
Circuits	84

CONTENTS.

SERIES ARC SYSTEM, DIRECT OR ALTERNATING CURRENT.

	PAGE
Dynamos	85
Foundations	87
Instruments	87
Cables to Switchboard	87
Arc Lamps	87
Hanger Boards	88
Hoods	88
Globes	89
Spark Arresters and Nets	89
Carbons	89
Starting Plant and Instruction	89
Renewal Parts	89
Summary	89
Switchboard Complete	90

FIXTURES, ETC.

Fixtures and Shades	91
Lamps	91
Sockets and Receptacles	92
Meters	92

INTERIOR WIRING. ALTERNATING OR DIRECT CURRENT, TWO-WIRE SYSTEM.

	PAGE
Outlets and Lights	94
System	94
Variation in Pressure	94
Insulation	95
Insulation Resistance	95
Subdivision of { Risers / Feeders }	95
Location of { Risers / Feeders }	96
Mains (If taps terminate in secondary distribution boxes)	96
Taps	96
Joints	97
Cut-out and Switch Cabinets	98
Cut-outs	99
Fuses	99
Switches	99
Fixture Supports	100
Meter Outlets	100
Elevator Lights	101
Molding	101

Additions and Deductions	101
Instruction	102
Hanging of Fixtures	102
Wiring and Attaching Sockets to Fixtures already in Place	103
Suspension of Sockets and Lamps . .	103

THREE-WIRE SYSTEM.

System	105

THREE-WIRE SYSTEM ADAPTED TO TWO-WIRE SYSTEM.

System	106

ARC SYSTEM.

Circuits	108
Insulation of Wire	108
Insulation Resistance	108
Joints	108
Method of Wiring	108
Suspension of Lamps	109

CONDUIT SYSTEM. TWO-WIRE.

Method of Wiring	110
Appliances	110
Placing of Conduits	110
Sealing Openings	110
Outlets and Lights	111
System	111
Variation in Pressure	111
Insulation	111
Insulation Resistance	111
Subdivision of {Risers, Feeders}	111
Location of {Risers, Feeders}	112
Mains (If taps terminate in secondary junction boxes)	112
Taps	113
Junction Boxes	113
Cut-outs	114
Fuses	114
Switches	114
Fixture Supports	114
Sizes of Tubes	114

CONTENTS.   xiii

	PAGE
Fastenings	114
Joints	114
Elbow Limitation	115
Outlets	115
Floor Work with Brass or Unarmored Conduits	115
Separation of Wires	115
Meter Outlets	116
Elevator Lights	116
Additions and Deductions	116
Instruction	116
Hanging of Fixtures	116
Wiring and Attaching Sockets to Fixtures already in Place	116
Suspension of Sockets and Lamps	116
Three-wire System	116
Three-wire System Adapted to the Two-wire System	116

INTERIOR WIRING FOR CENTRAL STATION PLANTS.

| Number of Lights | 117 |
| Fixtures | 117 |

CONTENTS.

	PAGE
Lamps	117
Sockets	117
System	117
Variation in Pressure	117
Insulation	118
Character of the Work	118
Circuits	118
Cut-outs and Switches	118
Meters	118
Suspension of Lamps	119
Molding	119
Additions and Deductions	119
Instruction	119
Hanging of Fixtures	119
Wiring and Attaching Sockets to Fixtures already in Place	119

POLE LINES. LOW POTENTIAL—DIRECT CURRENT SYSTEM—TWO- OR THREE-WIRE.

Franchises and Permits	120
Division of Circuits	121
Points of Control	121
Poles	121

	PAGE
Setting and Guying	121
Distances	122
Painting	122
Gains and Cross-arms	122
Pins and Insulators	123
Steps	123
Soil	123
Wiring	123
Joints	124
Lightning Arresters	125
Maximum Fall of Potential	125

ALTERNATING CURRENT SYSTEM.

Note	125

STREET LIGHTING CIRCUITS—ARC OR INCANDESCENT.

Note	126
Poles	126
Gains and Cross-arms	126
Steps	126
Suspension of Lamps	126

	PAGE
Fixtures (Incandescent)	127
Fixtures (Arc)	128
Wiring	129
Point of Control	129
Additions and Deductions	130
Instructions	130
Schedules	130

STORAGE BATTERY.

Type	137
Capacity	137
Number of Cells and Voltage	137
Erection	137
Connections	137
Guarantees	137
Summary	138

STEAM PLANT.

Note	139
Engines	139
Fittings	140

CONTENTS.

	PAGE
Painting	141
Foundations	142
Starting Plant and Instruction	143
Belts	143
Renewal Parts	144
Summary	144
Counter-shafting	145
Pulleys	145
Foundations	146
Boilers	146
Fittings	147
Setting	147
Stack	148
Smoke Connections	148
Piping	148
Condensers	150
Feed Pumps and Injectors	151
Feed-water Heaters and Purifiers	151
Separators	151
Gauge Board	151
Painting	152
Renewal Parts	152
Summary	153

RULES AND REQUIREMENTS OF THE
NATIONAL BOARD OF FIRE UNDER-
WRITERS 154
ARCHITECT'S AND BUILDER'S UNIFORM CON-
TRACT 204

INTRODUCTION.

The following "Specifications" are intended as outlines to aid in the construction of specifications for individual installations. They do not in any way take the place of the rules and regulations adopted by insurance companies and electric light associations, but on the contrary are so drawn as to depend largely on such rules for the details of construction work; such rules should therefore invariably be incorporated as provided for in the General Specification under "Inspection," Sec. 17, p. 40, and under "Insurance Rules," Sec. 18, p. 41.

For ready reference there is added to these specifications the latest Rules and Requirements of the National Board of Fire Underwriters governing electrical installations. We have also added the form of Uniform Contract adopted by the National Association of Builders and the American Institute of Architects.

INTRODUCTION.

It is not expected that these specifications will be followed in detail or without elaboration for all classes of work, and if so followed will lead into serious error, but they are presented with the hope that in their proper use constructing engineers, architects and others called upon to get out specifications may be relieved of some of the petty detail accompanying such work, which, though important, is too often neglected on account of its tediousness and the time it consumes.

Though it is expected that these specifications will be found sufficiently full and explicit for some of the smaller installations, not requiring special appliances or precautions, in the majority of instances a careful study must be made, for reasons which will suggest themselves, of the surrounding conditions and limitations and the specifications modified in accordance therewith.

No attempt has been made throughout the specifications to lay down ironclad rules or to volunteer technical information. It is fairly assumed that no one will attempt to make out a definite and detailed specification who is not himself competent to determine for any installation under consideration the best methods and system to be adopted and the kind and quality of materials and workmanship required by its purposes or by local conditions.

ELECTRIC LIGHTING
SPECIFICATIONS

ELECTRIC LIGHTING SPECIFICATIONS.

The outline given below is one that the author has often found convenient for use in checking a specification, or for reference in outlining the points to be taken up in detail in making up a specification.

1. Working outline.

In those cases where a formal specification is not required, as sometimes occurs in the case of a preliminary specification or a specification for a small isolated plant or wiring job, the necessary data for estimates and bids can often be easily and rapidly dictated, item by item, as suggested by these headings, with little liability of making any important omissions.

Plant to comprise?
- Dynamos and erection.
- Storage batteries and erection.
- Switchboard.
- Wiring { Interior. Exterior. }
- Fixtures.
- Pole line.
- Engines and erection.
- Boilers and erection.
- Piping and erection.

System?
- Incandescent.
  - Direct current.
  - Alternating current.
  - Constant current.
  - Constant potential.
  - High potential.
  - Low potential.
  - Series.
  - Two wire.
  - Three wire.
  - Combination.
- Arc. { Direct current. Alternating current. }
- Combination of arc and incandescent.

Dynamos?
- Number.
- Capacity in volts and amperes.
- Series.
- Shunt.
- Compound.
- Self-excited.
- Separately excited.
- Hand regulated.
- Automatically regulated.
- Direct-connected or belted.
- Foundation.
- Regulating instruments.
  - Pressure Regulators. { Hand. Automatic. }
  - Amperemeters.
  - Voltmeters.
  - Galvanometers.
  - Indicators.
  - Shunts.
  - Equalizers.
  - Compensators.
  - Impedance coils.
  - Synchronizers.
- Renewal parts.

ELECTRIC LIGHTING SPECIFICATIONS.

- **Storage Batteries?**
  - Type.
  - Maximum voltage.
  - Maximum amperes.
  - Total ampere-hours.
  - Kind of cells.
  - Erecting.
  - Connections.
  - Guarantees.

- **Wires to switch-boards?**
  - Insulation.
  - Wires to be run.
    - On insulators.
    - Under cleats.
    - In molding.
    - In conduit.
  - Capacity.

- **Switch-boards?**
  - Number.
  - Material.
  - Size.
  - Design.
  - To be wired how.
  - Instruments.
    - *See* dynamo regulating instruments.
    - Lightning arresters.
    - Ground detectors.
    - Testing sets.
    - Switches.
      - Single pole.
      - Double "
      - Three wire.
        - Plug.
        - Snap.
        - Knife.
        - Automatic.
        - Breakdown.
        - Throwover.
        - Dynamo changing.
        - Circuit changing.
        - Short circuiting.
        - Reversing, etc.
    - Cut-outs.
      - Single pole.
      - Double "
      - Three wire.
        - Fusible.
        - Automatic.
        - Magnetic.

Interior wiring ?
- Number of circuits.
- Number of outlets.
  - Lamp. { Incandescent. Arc. }
  - Switch.
  - Meter.
  - Motor.
- Number of lights. { Incandescent. Arc. }
- Insulation.
- Insulation resistance.
- Method of wiring.
  - Under cleats.
  - In molding.
  - On insulators.
  - In conduits.
- Variation in pressure.
- Cut-outs.
- Switches.
- Cabinets, junction boxes, etc.
- Meters. { Number. Kind. Capacity. }
- Converters. { Number. Capacity. }

Sockets and receptacles ?
- Number.
- Key.
- Keyless.
- Base. { Porcelain. Wood, etc. }
- Finish. { Plain. Polished. Plated, etc. }
- Waterproof.

ELECTRIC LIGHTING SPECIFICATIONS.

Lamps?
- Incandescent.
  - Number.
  - Voltage or amperage.
  - Candle-power.
  - Kind.
    - Plain.
    - Frosted.
    - Colored, etc.
- Arc.
  - Number.
  - Direct current.
  - Alternating current.
  - Series working.
  - Parallel "
  - Candle-power.
  - Single carbon.
  - Double "
  - Triple "
  - Plain or ornamental.

Fixtures?
- Incandescent.
  - Interior.
    - Number.
    - Kind.
    - Finish.
  - Exterior.
    - Hoods.
    - Reflectors.
    - Brackets.
    - Cross-suspension fixtures.
    - Waterproof globes.
- Arc.
  - Hoods.
  - Globes.
    - Plain.
    - Opal.
    - Ground.
    - Half-ground.
    - Colored, etc.
  - Spark arresters.
  - Nets.
  - Hanger boards.
  - Circuit cut-outs.
  - Outriggers.
  - Pole tops.
  - Mast arms.
  - Posts.
  - Lowering devices.

Pole line ?
- Number of circuits.
- Number of street lamps.
- Insulation.
- Poles.
- Cross arms.
- Pins and insulators.
- Steps.
- Line lightning arresters.
- Fall of potential.

Engines ?
- Number.
- Kind.
  - Horizontal.
  - Vertical.
  - High speed.
  - Medium speed.
  - Low speed.
- Direct connected.
- Belted.
- Horse-power.
- Single cylinder or compound.
- Condensing or non-condensing.
- Initial pressure.
- Back pressure.
- Speed.
- Fittings.
- Foundations.
- Belts.
- Renewal parts.

Counter-shafting ?
- Shafting.
- Pulleys and idlers.
- Clutches and couplings.
- Stands or hangers.
- Foundations.

Boilers ?
- Number.
- Kind.
  - Horizontal.
  - Vertical.
  - Return tubular.
  - Water tube.
- Horse-power.
- Fittings.
- Setting.
- Tools.
- Renewal parts.

ELECTRIC LIGHTING SPECIFICATIONS.

Stack ?
- Material.
- Diameter.
- Height.
- Flues and dampers.
- Pier or stack foundations.

Piping ?
- Method.
- Connection with heating system.
- Valves.
  - Ordinary gate or globe.
  - Reducing.
  - Back pressure.
- Atmospheric exhaust.
- Exhaust heads.
- Water supply, drips, blow off.
- Pipe covering.

Condensers ?
- Number.
- Kind.
- Capacity.

Pumps and injectors.
Heaters and purifiers.
Separators.
Oil extractors.
Date of commencement and completion
Date of starting plant.
Inspections.
Tests.
Terms of payment.

SPECIFICATION.

2. Warning. Parties making bids for any portion of the work contemplated under these specifications (and plans) must familiarize themselves therewith both as regards that portion of the work covered by their bid and such other work as must be carried on, or is intended to operate in conjunction therewith in order that the true spirit and intent of these specifications (and plans) may be fulfilled. In case these specifications (and plans) are in any part deficient or not clearly expressed, the parties making bids shall apply to ―――― for the required information before such bids are submitted, as no changes will be allowed in specifications (or plans) after the contract is awarded except under the conditions named in article "Additional, Omitted or Changed Work."

It must be understood and agreed that these specifications (and plans) shall be fulfilled in their true spirit and intent and that any apparatus or appliance essential to the proper and convenient operation of the system shall be supplied and installed without extra charge even though not specifically called for.

PREAMBLE.

3. Bids.

Parties bidding shall state specifically just what part of these specifications their bid covers.

A complete and correct copy of these specifications shall be attached to each bid submitted.

All bids must be submitted on or before ―――.

The right is reserved to reject any or all bids.

No bid will be considered unless accompanied by a certified check in the sum of ―――, payable to ―――, said check to be forfeited if the successful bidder shall fail to deposit with ――― within ――― days after the acceptance of his bid the bond required under these specifications. The checks of unsuccessful bidders will be returned to them within —— days from date of opening bids.

4. Bond. Successful bidders will be required to furnish an approved bond within —— days after the acceptance of their bid in the sum of —— to faithfully commence, carry on, and complete their work in every respect according to the true spirit and intent of these specifications.

5. Contractor. The word "contractor" as herein used refers to the party or parties whose bid or bids for the whole or any part of the work contemplated under these specifications have been accepted.

6. Commencement and Completion of Work. Contractors under these specifications shall commence work on dates to be assigned, notice to be given of such dates not less than —— days in advance. All work shall proceed as rapidly as is consistent with thoroughness and good workmanship, and shall be completed in the following times:

Installation of dynamos, storage batteries and apparatus within —— days after assigned date of commencement.

Installation of wiring and wiring devices within —— days after assigned date of commencement.

Erection of pole line and wire within —— days after assigned date of commencement.

Installation of fixtures and dependent work within —— days after assigned date of commencement.

Installation of steam plant complete within —— days after assigned date of commencement.

But delays due to strikes, riots, or accidents beyond the control of contractors shall be added to the time stipulated above for the completion of the work provided application is made in writing by the contractor at the time such delay occurs, giving its nature and extent, such application to be subject to the approval of ——.

7. Damages. If any contractor shall fail to complete his work in the time stipulated above, including time lost through unavoidable delays if such time has been approved, there shall be deducted as liquidated damages from the contract price the sum of ——— per day for each and every day the work remains uncompleted after the date set as above.

GENERAL SPECIFICATION.

8. Duties of Contractors. Each contractor shall personally or through an authorized and competent representative constantly supervise the work from its beginning to its completion and acceptance.

He shall, so far as possible, keep the same foremen and workmen on the work from its commencement to its completion and acceptance.

He shall furnish all transportation, labor, apparatus and materials necessary for performing his work according to the true spirit and intent of these specifications (and plans).

He shall obtain all necessary permits and licenses for temporary obstructions, etc., and shall pay all fees for same.

He shall at all times, until its completion and final acceptance, protect his work, apparatus and materials from accidental damage by other contractors or otherwise, making good any damage thus occurring at his own expense; also making good any injury done the building in the performance of his work.

He shall comply with all corporation, city, state and other ordinances and laws relating to his work.

He shall be responsible for all accidents resulting through his work.

He shall sub-let no portion of his work except on the written permission and approval of the ——— and shall be responsible for work thus sub-let as though it were his own.

(The purchaser) agrees to afford the contractor all reasonable facilities to enable the work to proceed without interruption from beginning to end and to make good any loss which the contractor suffers in consequence of delay on the part of said (purchaser.)

9. Work, Labor and Materials. All work contemplated under these specifications shall be executed in a workmanlike and substantial manner; no patched or slovenly work will be allowed.

The labor shall be thoroughly competent and skillful in its line.

All materials shall be of the very best quality, shall be of standard dimensions, unless specified otherwise, and samples shall be submitted to ——— and approved before being used.

10. Additional, Omitted or Changed Work. Additional work will be allowed only on the written order of (the purchaser.)

Specified work shall be omitted or changed only by written agreement between the contracting parties.

The addition or rebate for such added, omitted, or changed work shall be as mutually agreed upon, the amount to be stipulated in the order or agreement.

ELECTRIC LIGHTING SPECIFICATIONS.

11. Replacement of Defective Material.
The contractor shall make good for a period of —— days after the final acceptance of the work all defects which develop on account of defective work or material.

12. Patented Apparatus.
All patented apparatus and material must be furnished by the contractor under guarantee against loss through suits, royalties, or claims of any kind whatsoever, and that any loss or damage to (purchaser) through such suits or claims will be made good by said contractor.

13. Special Devices.
Every bidder is expected to include in his proposal not only everything called for in these specifications, but also any special devices or methods peculiar to his system which will add to the safety, completeness, or efficiency of the plant, stating clearly the advantages to be derived from their use.

14. Safeguards and Debris.
Contractors must provide all necessary safeguards from accidents to persons or

property; must keep all passages, entrances, sidewalks, etc., free from debris and incumbrances; and on the completion of the work must remove from the premises all surplus material of every kind and description.

15. Plans. All plans and detailed drawings necessary to show the scope and character of the work contemplated under these specifications will be furnished by the $\left\{ \begin{array}{l} \text{engineer} \\ \text{architect} \end{array} \right\}$ as required. Figured dimensions and detailed drawings are in all cases to be followed in preference to scaled dimensions. The interpretation of all plans and drawings shall rest with the $\left\{ \begin{array}{l} \text{engineer} \\ \text{architect} \end{array} \right\}$ and in case any doubt arises as to their interpretation or correctness, work shall be discontinued until such doubt is removed, or if continued it shall be at the risk of the contractor.

16. Tests. (Note.—The character and extent of

the tests, especially the final tests, must be determined for the most part by a consideration of each individual case. The purpose for which the plant is installed, unusual conditions to which any part may be subjected, necessary delays occurring during the process of construction, relation of one part of the installation to another, time intervening between the completion of the plant and its active operation, operation before completion and like considerations, should be given careful attention.)

All work shall be regularly and systematically tested while in process of construction and any defects found shall be immediately remedied.

The final tests shall be made in the presence of the $\{ \begin{array}{c} \text{engineer} \\ \text{architect} \end{array} \}$ or his representative, and the right is reserved by (the purchaser) in case any doubt arises as to the fulfillment of the true spirit and intent of the specifications, to demand a test by expert engineers selected as is usual in matters of arbitration, whose decision shall be final on all disputed points, the expense of such test to be borne equally by both parties unless the apparatus or material shall prove defective, in which case the contractor shall bear the expense and shall also remedy the defects. He shall also be liable for any damage or loss to (the purchaser) resulting from conditions incident to the remedying of such defects.

17. Inspection.   During its progress the work shall be subject to the inspection of the $\left\{ \begin{array}{c} \text{engineer} \\ \text{architect} \end{array} \right\}$ or his representative, and of the

$\left\{ \begin{array}{l} \text{———— Board of Fire Underwriters,} \\ \text{———— Board of Inspectors.} \end{array} \right\}$

On its completion a

$\left\{ \begin{array}{l} \text{Board of Fire Underwriters} \\ \text{Board of Inspectors} \end{array} \right\}$ certificate

shall be furnished (the purchaser) by the contractor stating that all the insurance rules and regulations under which the work was done have been complied with.

All costs of such inspection to be borne by the contractor.

This inspection is a part of the test and the work will not be considered ready for acceptance until the certificate has been delivered to ———.

18. Insurance Rules.
All work shall be done in accordance with the rules and regulations of ———.

19. Acceptance.
(Note.—The same considerations that determine the character of the tests will also enter largely into the conditions of the acceptance.)

(The purchaser) will assume no liability nor responsibility for any part of the installation until formally accepted in writing.

No part of the installation will be accepted until (the purchaser) is satisfied that it fully complies with the spirit and intent of the specifications.

The acceptance of any portion of the work shall not be construed as a final acceptance.

The final acceptance will be given only after the completion of the work contemplated under the specifications according to their true spirit and intent and after the final tests as specified. Such acceptance, however, shall not prejudice any claim which (the purchaser) may have for the replacement of defective material for the time specified.

The date of the completion of the final tests shall be taken as the date of such final acceptance provided such tests prove satisfactory.

20. Terms of Payment. (To conform to individual cases.)

INSTALLATION OF DYNAMOS AND SWITCH-BOARDS.

Low Potential, Direct Current System, Two-Wire or Three-Wire.

This contractor shall furnish, and, unless otherwise specified, erect the following apparatus and material:

21. Dynamos. (Note.—The following specification is for belted dynamos; for direct-connected, or direct-driven dynamos it must be modified in several particulars, but there is such considerable difference in the methods adopted by different manufacturers for adapting their dynamos for direct-connecting that it is difficult to frame a single specification to cover them all, and at best it would be an awkward affair. The principal points to be covered are the method of connection, such as by slipping the armature over the extended engine shaft or connecting engine and dynamo shafts by a coupling; the outboard bearing; the method of supporting the fields; the brush-holder support; the extended foundation box, and the speed.

It should also be mentioned where and by whom the dynamo is to be connected to the engine; how the freight and handling charges are to be divided; and where and by whom tests of the combination, if any are to be made before erection, are to be conducted.)

—— direct current, constant potential dynamo(s) (each) having a normal capacity of —— amperes at —— volts.

The (se) dynamo (s) shall be
{ shunt wound; }
{ compound wound; }
of the latest and most efficient pattern; mounted on a base provided with an adjustable belt tightener so that the belt may be tightened while in operation; capable of operating under full load for —— consecutive hours without increasing the temperature of any part, especial-

ly the armature, fields and commutator, to such a degree as to endanger the insulation or decrease the efficiency of operation; shall not spark appreciably with proper care of the commutator and adjustment of brushes, nor under considerable variation of load; shall have an insulation resistance of not less than—— ohms between all parts insulated from each other; shall be adapted to operate at such speed as will allow the use of high speed, automatic cut-off engines belted direct; shall be provided with efficient oiling devices; the armature shall be balanced both electrically and mechanically so that there will be no tendency to spring the shaft, or to draw the armature toward either bearing so as to cause excessive friction and heating, and no vibration; the dynamo(s) shall be so designed that with the proper connections any number may be operated in parallel of whatever ampere capacity, provided the voltage be the same; that when connected so to operate it shall be possible, with ordinary care and precaution, to add to or take away from the circuit any dynamo without in any manner affecting the operation of the remainder, or causing any change in the candle-power or steadiness of the lamps; and that when two or more are operating on the same circuit in parallel the load may be divided between them in proportion to their respective capacities under all conditions, from no load to full load; and that with the proper connections any number of pairs, a pair consisting of two similar dynamos, may be manipulated in the same manner and with the same effect as single dynamos, as indicated above.

The dynamo(s) shall be rated with such margin of safety that $\left\{ \begin{array}{c} \text{they} \\ \text{it} \end{array} \right\}$ shall not be injured if subjected to an overload of —— per cent. above such rating for a period of——.

22. Foundations.
(Note.—The following specification is for foundations for belted dynamos; while

the dynamo foundation need not be as massive as the engine foundation it should, nevertheless, be of ample dimensions.

For direct-connected outfits the foundation is the engine foundation sufficiently enlarged to also support the dynamo and therefore is properly included under the head of engine foundation.

For either purpose the foundation should be entirely separated from walls, floor or sub-stratum of rock by some deadening material such as felt, cork, or sand.)

The foundation(s) for the(se) dynamo(s) shall be built by the $\begin{Bmatrix} \text{contractor} \\ \text{purchaser} \end{Bmatrix}$ of —— laid ——, or other material subject to the approval of the $\begin{Bmatrix} \text{engineer,} \\ \text{architect,} \end{Bmatrix}$ and shall be of sufficient length, width and depth to safely and firmly sustain $\begin{Bmatrix} \text{their} \\ \text{its} \end{Bmatrix}$ weight.

$\begin{Bmatrix} \text{They} \\ \text{It} \end{Bmatrix}$ shall be insulated and isolated in the following manner:———, so that no vibration or noise whatsoever shall be transmitted to any part of the building.

Above the floor line $\left\{\begin{array}{c}\text{they}\\ \text{it}\end{array}\right\}$ shall be faced with ———.

All necessary excavating or filling and the removal of all debris shall be done by the $\left\{\begin{array}{c}\text{contractor}\\ \text{purchaser}\end{array}\right\}$ who shall also restore the floor in the following manner ———.

The height of the dynamo base-frame(s) above the floor line will be ———.

The dynamo base-frame(s) shall be firmly and securely fastened to the(se) foundation(s) in such a way as to prevent lateral motion in either direction, and to give an even bearing surface at every point.

23. Instruments. (Note.—This specification simply provides for such instruments as are actually necessary for operating the dynamo; if extra instruments are desired such as portable voltmeters, portable ammeters, testing sets, etc., they should be distinctly specified as well as the make if a particular kind is preferred.)

There shall be provided with each dynamo one hand regulator for adjusting the pressure, made entirely of incombustible material; one ampere meter for indicating the current supplied by the dynamo and graduated to read amperes; one voltmeter or pressure indicator, which may remain constantly in circuit so as to indicate continuously the pressure at the point to which it is connected, and which shall be so constructed that the scale is plainly visible at a distance of at least ——; one brush jig for trimming the brushes; and one insulating base frame provided with rails and a suitable device for shifting the position of the dynamo to alter the belt tension. There shall also be provided for the installation —— ground detector(s), which shall continuously indicate the insulation from the ground maintained on both sides throughout the system, and —— lighting arresters of the —— type.

24. Cables to Switchboard.
(Note.—In the case of compound dynamos operating in parallel the equalizing wire may often more conveniently be carried direct from dynamo to dynamo in which case this contractor should furnish and connect a suitable equalizing switch at each dynamo.)

This contractor shall carry to the switchboard location at —— all regulator and main wires, leaving the ends coiled up neatly, properly tagged, and sufficiently long to make the necessary switchboard connections. All main wires shall have a capacity of at least —— C. M. per ampere, and no wire smaller than —— B. & S. or —— B. W. G. shall be used. Regulator wires shall be covered with —— insulation, and shall be

$\left\{\begin{array}{l}\text{cleated to the ceiling,}\\ \text{carried on insulators,}\\ \text{concealed in molding,}\\ \text{run in conduits;}\end{array}\right\}$ main wires

shall be of $\left\{\begin{array}{l}\text{bare copper wire supported}\\ \text{on porcelain insulators,}\end{array}\right\}$

ELECTRIC LIGHTING SPECIFICATIONS.

or insulated wire,

{ carefully cleated to the ceiling,
carried on porcelain insulators,
concealed in molding consisting of a backing —— thick and a capping —— thick,
run in conduit, }

wires of opposite polarity being separated not less than ——. In no case shall insulated wires be carried in such proximity to heated surfaces, vapors or air as to endanger their insulation.

When the plant shall be ready for operation the dynamo(s) shall be operated for a period of —— consecutive days by competent engineers furnished by the contractor; all oil, waste, power, etc., to be furnished by (the purchaser). This contractor shall also give all necessary instruction to the engineer of (the purchaser) for the proper care, maintenance and operation of the dynamo(s), such instruction to be given during the trial period stipulated above.

25. Starting Plant and Instruction.

This contractor shall furnish such re-

26. Renewal Parts.

newal parts as it is advisable to keep on hand, adding hereto an itemized list of same.

27. Summary. (Note.—The object of this summary is to present in a concise, tabulated form, the essential data concerning the apparatus to be supplied, thus enabling the different bids to be easily and quickly compared. For complete plants a single form, systematically arranged, will be found extremely convenient both in making comparisons and in being able to see at a glance just what apparatus and appliances are called for, thus constituting a check on the specifications themselves.)

Each bidder shall fill out completely the following summary:

Number of dynamos . . . ———
Trade number or designation . ———
Rating in volts . . . . ———
Rating in amperes . . . ———
Shunt or compound . . . ———
Speed . . . . . ———
Dimensions of pulleys . . ———

H. P. required to be delivered at the pulley at full load . . . ———

Switchboard and Appliances.

This contractor shall furnish and erect the following apparatus and material:

(Note.—In plants of any considerable size the switchboard specification is an extremely important one. Its general characteristics will be determined by questions of purpose, economy, utility, available space, beauty, etc., but the details of material, method of wiring, attachment of instruments, location, number of switchboards, etc., require the most careful study. It may be advisable in one instance to have a single switchboard controlling everything from a single centre, in another to have a dynamo switchboard and a separate circuit switchboard, while in yet another to even divide the dynamo switchboard into two or more parts and to have several circuit switchboards. In determining the location of switchboards

28. Switchboard.

not only should questions of convenience be considered, but also questions of its relation to economy in the wiring.

The switchboard shall be made of ——, neatly and substantially built, of sufficient size to accommodate all the regulating apparatus, switches, bus bars, etc., named below without crowding, supported on a stout framing of ——, and set out not less than —— from the wall.

29. Switchboard Apparatus. (Note.—This specification includes only the instruments used in the simplest of installations. The instruments required for any particular installation must be determined by its individual purposes and necessities.)

There shall be placed upon this switchboard all the dynamo regulating apparatus and the following appliances:

——main ampere meters.
——ampere meters for risers.
——dynamo galvanometers.
——dynamo galvanometer switches.
——dynamo switches.

―― { riser / feeder } switches.

――change-over switches.

――break-down switches.

――main cut-outs.

(switches, cut-outs, etc., for purely local conditions).

All switches carrying over ―― amperes shall be knife switches; all other switches shall have sliding contacts, and shall make and break contact automatically beyond the control of the operator, who shall simply set the switch at the point of making or breaking.

Cut-outs shall be so protected that the molten metal cannot be spattered about on the fusing of the strip.

All switches and cut-outs shall be mounted on incombustible bases.

30. Connections (concealed). All connecting wires shall be carried back of the switchboard using only ―― wire. All joints shall be soldered. All connections to switches, cut-outs, etc., shall be soldered or made with an approved form of lug or set screw, in all cases care being taken to secure good and sufficient contact to prevent heating and insure permanency; when made with lugs or set screws they shall be in plain sight and easily accessible for tightening. Connecting wires shall be so run and secured that crosses or grounds are impossible in the normal operation of the plant. All main wires shall have a capacity of at least —— C. M. per ampere and no wire smaller than —— B. & S. or —— B. W. G. shall be used.

31. Connections (surface). All dynamo, bus, feeder and riser wires shall be of $\begin{Bmatrix} \text{bare} \\ \text{insulated} \end{Bmatrix}$ wire fastened neatly and securely to the front surface of the switchboard. All bare wires shall be separated from the board by an air space of not less than ——. Bus bars shall be of —— section. All minor connections, such as to pressure indicators, ground detectors, etc., shall be made on the $\begin{Bmatrix} \text{rear} \\ \text{surface} \end{Bmatrix}$ of the board using ——

wire. All joints shall be soldered. All connections to switches, cut-outs, etc., shall be soldered or made with an approved form of lug or set screw, in all cases care being taken to secure good and sufficient contact to prevent heating and insure permanency; when made with lugs or set screws they shall be in plain sight and easily accessible for tightening. All main wires shall have a capacity of at least —— C. M. per ampere, and no wire smaller than —— B. & S. or —— B. W. G. shall be used.

The following circuits will centre at the switchboard:

    (Enumeration of circuits
        To different floors;
        To different sections of the building;
        Residence circuits;
        Commercial circuits;
        Street lighting circuits;
        Power circuits, etc.

    This enumeration will largely determine the extra instruments to be supplied.)

32. Circuits.

Incandescent Series System, Dynamos Medium or High Potential (Variable or Constant), Current Direct or Alternating.

This contractor shall furnish, and, unless otherwise specified, erect the following apparatus and material:

33. Dynamos. (See Note: p. 43, sec. 21.)
―――― dynamo(s) (each) having a capacity of ―― kilowatts. The maximum voltage at the terminals of the dynamo(s) shall not exceed ―― volts at full load.

The dynamo(s) shall be of the latest and most efficient pattern; mounted on a base provided with an adjustable belt tightener so that the belt may be tightened while in operation; capable of operating under full load for ―― consecutive hours without increasing the temperature of any part, especially the armature, fields and commutator, to such a degree as to endanger the insulation or decrease the efficiency of operation; shall not spark unduly with proper care of the commutator and adjustment of the brushes, nor under considerable variation of load; shall have an insulation resistance of not less than —— ohms between all parts insulated from each other; shall be adapted to operate at such speed as will allow the use of high-speed, automatic cut-off engines belted direct; shall be provided with efficient oiling devices; the armature shall be balanced both electrically and mechanically so that there will be no tendency to spring the shaft or to draw the armature toward either bearing so as to cause excessive friction and heating, and no vibration; if of the direct and constant-current type $\left\{ \begin{matrix} \text{they} \\ \text{it} \end{matrix} \right\}$ shall be so designed that with the proper connections two or more may be operated successfully in series, so that with ordinary care and precaution any dynamo may be added to or taken from the circuit without in any manner affecting the efficient operation of the remainder, and with but a momentary fluctuation in the candle-power and steadiness of the lamps; if of the direct-current, constant-potential type $\begin{Bmatrix} \text{they} \\ \text{it} \end{Bmatrix}$ shall be so designed that with the proper connections two or more may be operated successfully in parallel so that with ordinary care and precaution any dynamo may be added to or taken from the circuit without in any manner affecting the operation of the dynamos remaining in circuit or causing any change in the candle-power of the lamps; and that when two or more are operating on the same circuit the load may be divided between them in proportion to their respective capacities under all conditions from no load to full load.

The dynamo(s) shall be rated with such margin of safety that $\begin{Bmatrix} \text{they} \\ \text{it} \end{Bmatrix}$ will not be injured if subjected to an overload of —— per cent. above such rating for a period of ——.

(See Note: p. 46, sec. 22.)

34. Foundation(s).

The foundation(s) for the(se) dynamo(s) shall be built by the { contractor / purchaser } of ——, laid ——, or of other material subject to the approval of the { engineer, / architect, } and shall be of sufficient length, width, and depth to safely and firmly sustain { their / its } weight. The foundation(s) shall be capped with a framing of well-seasoned timber securely fastened thereto, the dynamo baseframe(s) being securely fastened to the framing or to the foundation through the framing, both framing and baseframe being secured in such a manner as to prevent lateral motion in either direction and to give an even bearing surface at every point. If metal is used to fasten framing or baseframe to the foundation it must be thoroughly insulated where it passes through them, and at all places liable to come in contact with the dynamo must be countersunk and covered with a moisture-proof insulating compound.

All necessary excavating and filling and the removal of all debris shall be done by the {contractor / purchaser} who shall also restore the floor in the following manner: ———. The height of the dynamo base-frame(s) above the floor line will be——.

35. Instruments.

(See Note: p. 48, sec. 23.)

There shall be provided with each dynamo one voltmeter or pressure indicator which shall remain constantly in circuit so as to indicate at all times the pressure at the point to which it is connected; one brush jig for trimming the brushes; one insulating baseframe provided with rails and a suitable device for shifting the position of the dynamo to alter the belt tension; if of the constant-potential type, one hand regulator made entirely of incombustible material, a compensator or suitable balancing device for keeping the current in each circuit practically constant; if of the constant-current type a reg-

ELECTRIC LIGHTING SPECIFICATIONS.

ulator for so controlling the potential as lights are turned on or off that the current shall not vary appreciably from its normal value under any condition of load. There shall be provided for each circuit one ampere meter for indicating the current in said circuit, one ground detector which shall continuously indicate the insulation from the ground maintained at both poles and approximately the distance of any ground from the station, and one pair of lighting arresters. There shall also be provided one testing set capable of measuring up to —— ohms.

36. Cables to Switchboard. This contractor shall carry to the switchboard location at —— all regulator, exciter, auxiliary and main wires, leaving the ends coiled up neatly, properly tagged and sufficiently long to make the necessary switchboard connections. All main wires shall have a capacity of at least —— C. M. per ampere and no wire smaller than —— B. & S. or —— B. W. G. shall be used. All wires shall be insulated with ——; shall be run ——; when of opposite polarity shall be separated at least ——; and where crossing each other, wires of other circuits, or passing near metal pipes, girders, etc., shall be further protected by ——. In no case shall insulated wires be carried in such proximity to heated surfaces, vapors or air as to endanger their insulation.

37. Starting Plant and Instruction. (See p. 51, sec. 25.)

38. Renewal Parts. (See p. 51, sec. 26.)

39. Summary. (See Note: p. 52, sec. 27.)
Each bidder shall fill out completely the following summary:
Number of dynamos . . . . . ——
Trade number or designation . ——
Rating in volts . . . . ——
Rating in amperes . . . . . ——
Direct or alternating current . . ——
Constant or varying current . . ——
Constant or varying potential . . ——
Series, shunt, compound, self or

ELECTRIC LIGHTING SPECIFICATIONS.

    separately excited . . . . . ———
Rating of exciter in volts . . . ———
Rating of exciter in amperes . . ———
Trade number or designation . . ———
Regulation by hand or automatic ———
Number of fully loaded circuits pos-
    sible to operate per dynamo . ———
Volts per —— c. p. lamp . . . ———
Amperes per —— c. p. lamp . . ———
H. P. required to be delivered at
    the pulley at full load . . . ———

(See p. 53, sec. 28.)

40. Switchboard.

There shall be placed upon this switchboard, in addition to all the dynamo regulating apparatus, such switches, cut-outs and other appliances as are necessary for the proper and convenient manipulation of the circuits, such appliances to be named by each bidder in his proposal. For systems operating dynamos and circuits in parallel, the appliances and connections shall be such as will permit adding to or taking from the circuit any

41. Switchboard Apparatus and Connections.

dynamo without in any manner affecting the operation of the dynamos remaining in circuit or the candle-power of the lamps, and will permit the cutting in or out of any circuit without affecting the stability of other circuits. For systems operating one dynamo for each circuit or series of circuits the connections shall be such as will permit any circuit to be connected to or disconnected from any dynamo with certainty and rapidity. (For switchboard connections see secs. 30 and 31, pp. 55 and 56.

42. Circuits. (See p. 57, sec. 32.)

Constant Potential, Alternating Current System.

This contractor shall furnish, and, unless otherwise specified, erect the following apparatus and material:

(See Note: p. 43, sec. 21.)   43. Dynamo(s).
―――― constant potential, alternating current dynamo(s) (each) having a capacity of ―――― kilowatts. The maximum voltage at the terminals of the dynamo(s) shall not exceed ―――― volts.

The dynamo(s) shall be of the latest and most efficient pattern; mounted on a base provided with an adjustable belt tightener so that the belt may be tightened while in operation; capable of operating under full load for ―――― consecutive hours without increasing the temperature of any part, especially the armature, fields and commutator, to such a degree as to endanger the insulation or decrease the efficiency of operation; shall have an insulation resistance of not less than —— ohms between all parts insulated from each other; shall be adapted to operate at such speed as will allow the use of high-speed, automatic cut-off engines belted direct; shall be supplied with efficient oiling devices; the armature shall be balanced both electrically and mechanically so that there will be no tendency to spring the shaft, or to draw the armature toward either bearing so as to cause excessive friction and heating, and no vibration; all contacts, brushes, binding posts, etc., shall be so placed and protected that there is the least possible danger of receiving a shock. The dynamo(s) shall be rated with such margin of safety that $\left\{ \begin{array}{l} \text{they} \\ \text{it} \end{array} \right\}$ will not be injured if subjected to an overload of —— per cent. above such rating for a period of ——; if self-exciting the coils furnishing the exciting current, and the commutator shall be so insulated and protected that it will be impossible under the ordinary conditions of operation to ground or cross them on themselves or on the armature circuit; if separately excited the exciting dynamo(s) also shall be subject to the general conditions given above, shall operate at a potential not exceeding —— volts, shall be of ample capacity to excite the fields of —— dynamo(s) (each) having a capacity of —— kilowatts, and shall operate with no appreciable sparking at the brushes.

(See p. 61, sec. 34.)

44. Foundations.

(See Note: p. 48, sec. 23.)

45. Instruments.

There shall be provided with each dynamo one ampere meter for indicating the current supplied by the dynamo; one voltmeter or pressure indicator which shall remain constantly in circuit so as to indicate continuously the pressure on the primary mains at the point to which it is connected, and shall be so constructed that the scale is plainly visible at a distance of at least——;one insulating baseframe provided with rails and a suitable device for shifting the position of the dynamo to alter the belt tension; one ground detector which shall continuously indicate the insulation maintained on both sides throughout the system; one pair of lightning arresters; for the field-exciting circuit one hand regulator made entirely of incombustible material, one voltmeter, one brush jig, one double-pole knife switch mounted on an incombustible base, one double-pole fusible cut-out mounted on an incombustible base; if the fields are separately excited there shall be provided with the exciting dynamo one insulating baseframe as above, one endless belt, and also, if it excites the fields of more than one dynamo, one hand regulator for the field circuit of the exciting dynamo, one hand regulator for each of the field circuits of the excited dynamos, and one double-pole knife switch and double-pole fusible cut-out for each of the above field circuits, including that of the exciting dynamo unless lights are operated therefrom, in which case a switch and cut-out shall be placed in the main circuit before branching off to the various dynamos.

There shall be provided the following numbers and sizes of converters: 46. Converters.

Number.	Size.
———	——— lt.
etc.	etc.

Each converter shall reduce the voltage on the primary circuit to —— volts on the secondary circuit; shall have its capacity plainly marked upon it; and shall be provided with a separate fuse-box which shall be so arranged that when replacing fuses or otherwise working about the converter the primary circuit may be opened.

(See p. 63, sec. 36.) 47. Cables to Switchboard.

(See p. 51, sec. 25.) 48. Starting Plant and Instruction.

(See p. 51, sec. 26.) 49. Renewal Parts.

50. Summary. (See Note: p. 52, sec. 27.)

Each bidder shall fill out completely the following summary:

Number of alternators . . . . . ———
Trade number or designation . . ———
Number of exciters . . . . . ———
Trade number or designation . . ———
Rating of alternators in volts . . ———
Rating of alternators in amperes . ———
Rating of exciters in volts . . . ———
Rating of exciters in amperes . . ———
Self or separately excited . . . ———
Reduction recommended in converters . . . . ———
H. P. required to be delivered at the pulley at full load . . . ———

51. Switchboard. (See p. 53, sec. 28.)

52. Switchboard Apparatus and Connections. There shall be placed upon this switchboard, in addition to the dynamo regulating apparatus, such switches, cut-outs and other appliances as are necessary for the proper and convenient manipulation of the circuits, such appliances to be

ELECTRIC LIGHTING SPECIFICATIONS. 73

named by each bidder in his proposal.

The appliances and connections shall be such as will permit the operation of any dynamo on any circuit or number of circuits, and the cutting in or out of any dynamo or circuit with certainty and rapidity without in any manner affecting the operation of other dynamos or circuits. (They shall also be arranged in such manner as to render it possible to connect two dynamos in parallel by an arrangement of switches or combination of circuits.) (For switchboard connections see secs. 30 and 31, pp. 55 and 56.)

(See p. 57, sec. 32.)   53. Circuits.

Alternating Current or Direct Current System, with the Parallel System of Distribution.

This contractor shall furnish, and, unless otherwise specified, erect the following apparatus and material:

54. Dynamo(s). (See Note: p. 43, sec. 21.)
A dynamo capacity of, as nearly as possible, —— kilowatts. (The number of dynamos shall not be less than ——, nor more than ——.)

The dynamo (s) shall be of the latest and most efficient pattern; shall be mounted upon a base provided with an adjustable belt tightener so that the belt may be tightened while in operation; capable of operating under full load for ——consecutive hours without increasing the temperature of any part, especially the armature, fields, and commutator, to such a degree as to endanger the insulation or decrease the efficiency of operation; shall not spark appreciably with proper care of the commutator and adjustment of the brushes, nor under considerable variation of the load; shall have an insulation resistance of not less than ——ohms between all parts insulated from each other; shall be adapted to operate at such speed as will allow the use of high-speed, automatic cut-off engines belted direct; shall be provided with efficient oiling devices; the armature shall be balanced electrically and mechanically so that there will be no tendency to spring the shaft, or to draw the armature toward either bearing so as to cause excessive friction and heating, and no vibration; all contacts, binding posts, brushes, etc., having considerable differences of potential between them shall be so placed and protected that the danger of receiving a shock is the least possible; armature conductors shall be so securely attached to the armature as to preclude any possibility of their being dragged from their proper position under normal conditions of operation; all coils, connections, commutators, brushes, etc., belonging to different circuits shall be so insulated and protected that it will be impossible under the ordinary conditions of operation to cross or ground them; the dynamo(s) shall be rated with such margin of safety that $\begin{Bmatrix} \text{they} \\ \text{it} \end{Bmatrix}$ will not be injured if subjected to an over-load of —— per cent. above $\begin{Bmatrix} \text{their} \\ \text{its} \end{Bmatrix}$ normal rating for a period of ——; if (an) auxiliary dynamo(s) $\begin{Bmatrix} \text{are} \\ \text{is} \end{Bmatrix}$ required for the operation of the dynamo(s) specified above $\begin{Bmatrix} \text{they} \\ \text{it} \end{Bmatrix}$ shall be subject to the same conditions.

55. Foundations. (See Note: p. 46, sec. 22.)

The foundation(s) for the(se) dynamo(s) shall be built by the $\begin{Bmatrix} \text{contractor} \\ \text{purchaser} \end{Bmatrix}$ of ——, laid ——, or of other material subject to the approval of the $\begin{Bmatrix} \text{engineer,} \\ \text{architect,} \end{Bmatrix}$ and shall be of sufficient length, width and depth to safely and firmly sustain $\begin{Bmatrix} \text{their} \\ \text{its} \end{Bmatrix}$ weight.

For dynamos operating at voltages below ―― the baseframe may be firmly and securely fastened directly on the foundation. For dynamos operating at voltages above ―― the foundation shall be capped with a framing of well-seasoned timber securely fastened thereto, the baseframe being securely fastened to the framing or to the foundation through the framing; both framing and baseframe being secured in such a manner as to prevent lateral motion in either direction, and to give an even bearing surface at every point.

If metal is used to fasten framing or baseframe to the foundation it shall be thoroughly insulated where it passes through them and at all places liable to come in contact with the dynamo shall be countersunk and covered with a moisture-proof insulating compound.

All necessary excavating and filling, and the removal of all debris shall be done by the { contractor / purchaser } who shall also restore the floor in the following manner: ——— ———.

The height of the dynamo base-frame(s) above the floor line will be ———.

56. Instruments.

(See Note: p. 48, sec. 23.)

There shall be provided with each dynamo one hand regulator for adjusting the pressure, made entirely of incombustible material; one ampere meter for indicating the current supplied by the dynamo; one voltmeter or pressure-indicator which shall remain constantly in circuit so as to indicate continuously the pressure at the point to which it is connected, and shall be so constructed that the scale is plainly visible at a distance of at least ———; one insulating base-frame provided with rails and a suitable device for shifting the position of the dynamo to alter the belt tension; there shall be provided for the installation one ground detector, or if the circuits have no common junction one ground detector for each circuit which shall continuously indicate the insulation from the ground maintained on both sides throughout the system, and one pair of lightning arresters for the common circuits, or for each circuit if they are kept separate; there shall also be provided such brush jigs, switches, cut-outs, belts and other appliances as are requisite and proper for the operation of the system; all switches and cut-outs to be of the double pole type and mounted on incombustible bases.

57. Converters.

For alternating systems there shall be provided the following numbers and sizes of converters:

Number.	Sizes.
———	———lt.
etc.	etc.

(See Note: p. 50, sec. 24.)

58. Cables to Switchboard.

This contractor shall carry to the switchboard location at —— all regulator, exciter, auxiliary, and main wires, leaving the ends coiled up neatly, properly tagged and sufficiently long to make the necessary switchboard connections. All main wires shall have a capacity of at least —— C. M. per ampere, and no wire smaller than —— B. & S. or ——B. W. G. shall be used. All regulator, exciter, and for systems carrying over —— volts all auxiliary and main wires, shall be insulated with ——; all wires of systems carrying over —— volts shall be run ——; where of opposite polarity shall be separated at least ——; and where crossing each other, wires of other circuits or passing near metal pipes, girders, etc., shall be further protected by ——.

For systems carrying less than —— volts the main wires shall be of
{ bare copper wire supported on porcelain insulators.

ELECTRIC LIGHTING SPECIFICATIONS.

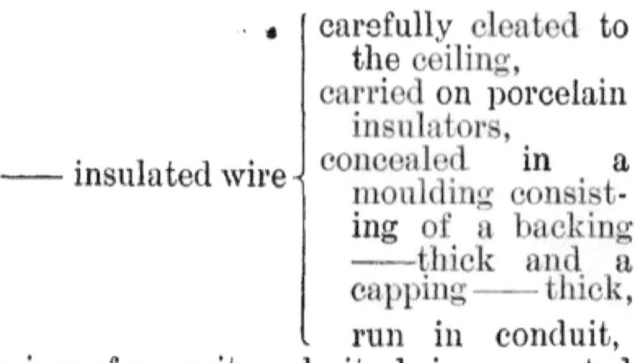

—— insulated wire { carefully cleated to the ceiling, carried on porcelain insulators, concealed in a moulding consisting of a backing ——thick and a capping —— thick, run in conduit,
wires of opposite polarity being separated not less than ——.

In no case shall insulated wires be carried in such proximity to heated surfaces, vapors or air as to endanger their insulation.

(See p. 51, sec. 25.)   59. Starting Plant and Instruction.

(See p. 51, sec. 26.)   60. Renewal Parts.

(See Note: p. 52, sec. 27.)   61. Summary.

Each bidder shall fill out completely, so far as it pertains to his apparatus, the following summary:

Number of direct-current dynamos ——
Trade number or designation . . ——

Number of alternating-current
   dynamos . . . . ———
Trade number or designation . . ———
Number of exciter dynamos . . ———
Trade number or designation . . ———
Rating of dynamos in volts . . ———
Rating of dynamos in amperes . ———
Rating of exciters in volts . . . ———
Rating of exciters in amperes . . ———
Self or separately excited . . . ———
Shunt or compound . . . . . ———
Reduction recommended in convert-
   ers . . . . ———
H. P. required to be delivered at pulley
   at full load . . . . . . ———

62. Switch-board. (See p. 53, sec. 28.)

63. Switch-board Apparatus and Connections. There shall be placed upon this switch-board, in addition to all the dynamo regulating apparatus, such switches, cut-outs and other appliances as are necessary for the proper and convenient manipulation of the circuits, such appliances to be named by each bidder in his proposal.

For the two-wire, direct-current system the appliances and connections shall be such as will permit any dynamo to be added to or taken from parallel circuit without in any manner affecting the operation of the dynamos remaining in circuit, or the candle-power and steadiness of the lamps.

For the three-wire, direct-current system the appliances and connections shall be such as will permit any dynamo on either side to be added to or taken from parallel circuit without in any manner affecting the operation of the dynamos remaining in circuit on that side, or the candle-power and steadiness of the lamps; and a breakdown switch for connecting the two outside wires in case it shall become necessary or desirable to operate on the two-wire system.

For the alternating-current system the appliances and connections shall be such as will permit the operation of any dynamo on any circuit or number of circuits, and the cutting in or out of any dynamo or circuit with certainty and rapidity, without in any manner affecting the operation of other dynamos or circuits (and such that it will be possible to connect two dynamos together by an arrangement of switches or combination of circuits.) (For switchboard connections see secs. 30 and 31, pp. 55 and 56.

64. Circuits.   (See p. 57, sec. 32.)

Series Arc System, Direct or Alternating Current.

This contractor shall furnish, and, unless otherwise specified, install the following apparatus and material:

―― arc light dynamo(s) (each) having a capacity of ――, ―― c. p., ―― ampere, ―― volt arc lamps. 65. Dynamo(s).

Each dynamo shall be provided with a regulator which shall automatically make the proper adjustments for all changes of load from no load to full load, the adjustments to be made in such a way as not to endanger any part of the dynamo, appliances or lamps, nor to cause any perceptible change in the balance remaining in operation; shall be of the latest and most efficient pattern; mounted on a base provided with an adjustable belt-tightener, so that the belt may be tightened while in operation; capable of operating under full load for —— consecutive hours without increasing the temperature of any part, especially the armature, fields, and commutator, to such a degree as to endanger the insulation or decrease the efficiency of operation; shall have an insulation resistance of not less than —— ohms between all parts insulated from each other; shall be adapted to operate at such speed as will allow the use of high-speed, automatic-cut-off engines belted direct; shall be provided with efficient oiling devices; the armature shall be balanced both electrically and magnetically so that there will be no tendency to spring the shaft or to draw the armature toward either bearing so as to cause excessive friction and heating, and no vibration; especial attention shall be given the insulation, protection and separation of contacts, binding posts and bared surfaces having extreme differences of potential in order to minimize the danger of accidental shocks, crosses, or grounds under normal conditions of operation; the dynamo(s) shall be so designed and automatically regulated that the power will be automatically proportioned to the number of lamps burning at any time.

(See p. 61, sec. 34.)

66. Foundations.

(See Note: p. 48, sec. 23.)

67. Instruments.

There shall be provided with each dynamo, in addition to the automatic regulator required above, one ampere meter for indicating the current supplied by the dynamo and graduated to read amperes; one brush jig for trimming the brushes; one insulating baseframe, provided with rails and a suitable device for shifting the position of the dynamo to alter the belt tension; one main switch; one pair of lightning arresters; and for the general installation one testing set capable of measuring up to —— ohms.

(See p. 63, sec. 36.)

68. Cables to Switchboard.

There shall be provided

69. Arc Lamps.

——— 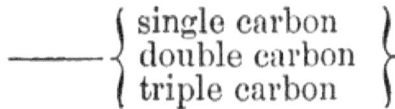 arc lamps of ——— nominal candle-power. Each lamp shall be provided with a switch by which it may be cut in or out of circuit; shall be regular in its feeding action; shall be free from hissing, flickering or flaming when provided with proper carbons; shall contain an efficient device which shall automatically cut out a lamp for any reason defective, without interfering with the operation of the lamps remaining in circuit; and shall be simple, strong, and durable in its mechanical construction.

70. Hanger Boards. ——— hanger boards for inside use, each hanger board to contain a switch by which the lamp may be cut entirely out of circuit.

71. Hoods. ——— waterproof hoods, complete with { hanger boards for outside use / out-rigger attachments / cross-suspension attachments }

ELECTRIC LIGHTING SPECIFICATIONS.

——— { clear glass / full ground / half ground / opal / colored, etc. } globes.   72. Globes.

——— wire-gauze spark arresters, and ——— wire globe nets.   73. Spark Arresters and Nets.

——— sets of carbons, a set consisting of one upper and one lower carbon.   74. Carbons.

(See p. 51, sec. 25.)   75. Starting Plant and Instruction.

(See p. 51, sec. 26.)   76. Renewal Parts.

(See Note: p. 52, sec. 27.)   77. Summary.

Each bidder shall fill out completely the following summary:

Number of dynamos . . . . ———
Trade number or designation . . ———
Rating in volts . . . . ———
Rating in amperes . . . . . ———
Capacity in ——— c. p. lamps . . ———
Series or shunt wound . , . . ———
H. P. required to be delivered at pulley at full load . . . ———

ELECTRIC LIGHTING SPECIFICATIONS.

SWITCHBOARD APPARATUS AND CONNECTIONS.

This contractor shall furnish and erect the following apparatus and material:

78. Switchboard Complete.
One combination (material) switchboard with a capacity for —— circuits, and provided with the necessary sockets, plugs, main and transfer cables, testing connections, and a suitable and convenient device for holding cables not in use. It shall be so arranged and marked that any circuit or series of circuits may be quickly connected with or disconnected from any dynamo with the least possible danger of short-circuits or error. Sockets shall be so designed that it is practically impossible to short-circuit, ground or receive a shock from them. All connections with the dynamo leads shall be easily accessible. All wires used in making connections shall have —— insulation. All plugs shall have well insulated wooden handles and the cables shall be covered with soft-rubber tubing

or equivalent as an extra precaution. All cables shall be of stranded wire.

All joints shall be soldered. All connections to switches, cut-outs, etc., shall be soldered or made with an approved form of lug or set-screw, in all cases care being taken to secure good and sufficient contact to prevent heating and insure permanency; when made with lugs or set-screws they shall be in plain sight and easily accessible for tightening. Connecting wires shall be so run and secured that crosses or grounds are impossible in the normal operation of the plant.

FIXTURES, ETC.

(Note.—No set specification can be made for fixtures and shades; their character must be determined wholly from individual requirements. See schedule, p. 136). _{79. Fixtures and Shades.}

This contractor shall furnish and deliver at —— the following number, sizes, and kinds of incandescent lamps: _{80. Lamps.}

Number.	C. p.	Voltage.		Plain.
—	—	Amperage		Frosted.
etc.	etc.	—		Colored, etc.
		etc.		—
				etc.

stating, also, their make and the make of the socket for which they are adapted.

Lamps shall be guaranteed to have an average life of not less than —— hours if burned at their normal voltage. They shall burn with a white light and shall not blacken under proper use. All lamps giving out or proving defective during the trial period of —— days under normal and proper use shall be replaced without charge.

81. Sockets and Receptacles. This contractor shall furnish and deliver at —— the following numbers and kinds of sockets and receptacles:

Number.	Kind.	Finish.
etc.	etc.	etc.

82. Meters. This contractor shall furnish and $\begin{Bmatrix} \text{deliver at} \text{——} \\ \text{place in position} \end{Bmatrix}$ the following numbers and sizes of the ——

ELECTRIC LIGHTING SPECIFICATIONS.

{ watt meter,
  recording ampere meter,
  current counter.

Number.	Capacity in amperes.	Two or three wire.
etc.	etc.	etc.

INTERIOR WIRING.

Alternating or Direct Current, Two-Wire System.

83. Outlets and Lights. The building shall be wired to —— lamp outlets, —— switch outlets, (and —— meter outlets) for the equivalent of ——, —— ampere, —— volt lamps. The wiring shall be (to outlets only) (except for cut-outs and switches; cut-outs and switches shall be furnished and installed complete). At each outlet the loose wire shall be neatly coiled and the ends carefully taped.

84. System. All wiring shall be for the parallel two-wire system of distribution.

85. Variation in Pressure. The fall of potential between the switchboard (centre of distribution) and the (farthest lamp) shall not exceed at

—— load —— per cent. of the initial pressure; this difference to be divided as follows: { risers / feeders } —— per cent., mains —— per cent., taps —— per cent.

86. Insulation. All wire used throughout the installation shall be insulated with ——.

87. Insulation Resistance. Each { riser, / feeder, } main, and tap shall test out with an insulation resistance of at least —— ohms.

88. Subdivision of {Risers} {Feeders}. From the switchboard (or centre of distribution) —— { risers, / groups of risers, / feeders, } shall be carried to the following points: { Riser / Group / Feeder } No. 1 to ——, { riser / group / feeder } No. 2 to ——, etc.

{ Riser / Group / Feeder } No. 1 shall feed all lights (location), $\begin{Bmatrix} \text{riser} \\ \text{group} \\ \text{feeder} \end{Bmatrix}$ No. 2 shall feed, etc.

89. Location of {Risers} {Feeders.}

From the switchboard (or centre of distribution) the $\begin{Bmatrix} \text{risers} \\ \text{feeders} \end{Bmatrix}$ shall be carried $\begin{Bmatrix} \text{under cleats,} \\ \text{on insulators,} \\ \text{in moulding,} \\ \text{in conduit, etc.,} \end{Bmatrix}$ to ——, and thence upward in (channels, wooden conduits, elevator shaft, air shaft, etc., with location) to their respective cut-out boxes.

90. Mains (If taps terminate in secondary distribution boxes).

From the $\begin{Bmatrix} \text{riser} \\ \text{feeder} \end{Bmatrix}$ cut-out boxes mains shall be carried $\begin{Bmatrix} \text{under cleats,} \\ \text{on insulators,} \\ \text{in molding,} \\ \text{in conduit, etc.,} \end{Bmatrix}$ to —— secondary cut-out boxes where all tap lines shall centre. From $\begin{Bmatrix} \text{riser} \\ \text{feeder} \end{Bmatrix}$ No. 1 shall be carried —— mains terminating at ——; from $\begin{Bmatrix} \text{riser} \\ \text{feeder} \end{Bmatrix}$ No. 2, etc.

91. Taps.

(Note.—In certain cases it is advisable

to run the circuits in such a manner that no room shall be dependent on one circuit only; if so desired it should be added under this heading.)

From the $\begin{Bmatrix} \text{riser} \\ \text{feeder} \\ \text{main} \end{Bmatrix}$ cut-out boxes distributing circuits shall be run to the various outlets as specified in the schedule and located on the plans. The wires shall be run $\begin{Bmatrix} \text{under cleats,} \\ \text{on insulators,} \\ \text{in molding,} \\ \text{in conduit, etc.,} \end{Bmatrix}$ in such a manner that the highest possible insulation shall be maintained under all circumstances.

Except in case of single outlets for a group of lamps and circuits specifically mentioned no distributing circuit shall carry over —— amperes. Distributing circuits shall be of one size of wire throughout their entire length.

92. Joints.

Throughout the installation joints shall be avoided where possible; where absolutely necessary they must be made mechanically strong and secure, carefully soldered, wiped free from any moisture and excess of flux and so taped and compounded that the insulation of the joints shall be equal to the original insulation; the solder shall be relied on only to give a good electrical connection.

93. Cut-Out and Switch Cabinets. (Note.—A complete description of the cabinets should be given covering material, doors, hinges, locks, finish, etc.; also stating what parts, if any, will be furnished by other contractors, specifying what switches are to be placed in the cabinets, and locating and describing cabinets for switches alone if such are to be provided; if name plates are to be furnished so specify and describe.)

The terminals of all $\begin{Bmatrix} \text{risers,} \\ \text{feeders,} \end{Bmatrix}$ mains and taps shall be brought together in cabinets at the points designated in this specification and on the plans. (If wiring contractor does not furnish cut-outs and switches, add: •The terminals of the $\begin{Bmatrix} \text{risers,} \\ \text{feeders,} \end{Bmatrix}$ mains and taps shall be brought into these cabinets in such a way as to permit the easy and convenient insertion and connection of the cut-outs and switches specified.)

94. Cut-Outs. A cut-out shall be provided for each branch circuit. All cut-outs shall be double-pole, mounted on incombustible bases, and with connections of such size and shape as to afford ample contact surface for both conductors and fuses.

95. Fuses No fuses shall be put in the cut-outs except by special order, but a complete supply, consisting of not less than —— sets for each cut-out, shall be provided. These fuses shall be of the plug type or furnished with metal tips and shall have their capacity plainly marked upon them.

96. Switches. For the number and capacity of switches see the attached schedule.

All switches shall be double-pole, mounted on incombustible bases, with automatic make and break, the switch being merely set at the point of making and breaking by the operator, and with sliding contacts. The capacity of each switch shall be plainly marked upon it and shall not be less than —— per —— c. p. lamp controlled.

97. Fixture Supports. Where no fixture support is provided this contractor shall furnish for all $\left\{ \begin{array}{l} \text{side} \\ \text{ceiling} \end{array} \right\}$ outlets a suitable support consisting of a wooden block firmly fastened to the wall flush with the plaster, and of sufficient dimensions to securely hold the fixture, a piece of gas-pipe securely anchored by means of an iron plate; or such other device as shall be best adapted to the construction of the building and the character of the fixture to be installed in each particular location.

(Specify approximate number required.)

98. Meter Outlets. At the places located in the schedule and on the plans, meter outlets shall be run (and a support for each meter provided consisting of —— securely fastened to the wall).

Each elevator is to be provided with ——, —— c. p. lamp(s). Each elevator shall be on its own cut-out and circuit which shall be run from the distribution box at (location). The wiring shall include the wiring of the elevator car, all necessary cables and the connection with its outlet. The cables shall be well insulated, flexible and properly protected from abrasion. 99. Elevator Lights.

The molding used in the places specified above shall be of ——, and finished ——. 100. Molding.

On all outside walls, bare brick or stone walls, etc., it shall consist of a backing and capping.

While the schedule is intended to represent very closely the number of lights 101. Additions and Deductions.

and outlets to be wired to, yet, as some changes may become necessary during the process of construction, each bidder shall name in his proposal a price to be added to or deducted from the contract price for each light or outlet wired for in excess of the number specified, or which shall be cancelled, provided such addition or cancellation involves no change in the work already completed and shall be along the lines of existing circuits.

102. Instruction. (Note.—It may be desirable that the purchaser furnish one man to work under the contractor in order that he may have a man thoroughly familiar with all the details of the construction; the contractor to give such instruction as will enable him to acquire a thorough and intelligent knowledge of methods, appliances, location of circuits, etc.)

103. Hanging of Fixtures. (Note.—To be inserted if this contractor is to complete the wiring, including the attachment of lamps and sockets.)

This contractor shall hang all fixtures, including the assembling and wiring of the fixtures (unless provided for under Fixtures), the attaching of sockets, lamps and shades, and the connection with the ends of the taps.

Insulating joints will be furnished, where required, by the fixture contractor but this contractor shall furnish and connect a suitable cut-out for each outlet, protecting both sides of the circuit.

104. Wiring and Attaching Sockets to Fixtures already in Place.

This contractor shall attach —— sockets by means of a (suitable gas attachment) to the (gas) fixtures already in place. These fixtures shall be properly insulated and shall be wired in the following manner: —— ——, with ——. At each outlet a suitable cut-out shall be provided protecting both sides of the circuit. In each socket the proper lamp shall be placed and all shades shall be attached.

105. Suspension of Sockets and Lamps.

Sockets and lamps shall be suspended by means of —— pendants from the ceiling. Each pendant to be —— in length, (provided with a cord adjuster) and protected by a double-pole ceiling cut-out. Both at the cut-out and in the socket the cord shall be knotted so that in no case will the weight come on the binding screws. Where the cord passes through the neck of the socket it shall be protected by a —— bushing. In each socket the proper lamp shall be placed and all shades shall be attached.

Three-Wire System.

This specification is identical with the two-wire specification, except in "System," p. 94, sec. 84, in place of which insert the following:

All $\begin{Bmatrix} \text{risers} \\ \text{feeders} \end{Bmatrix}$ and mains shall be figured on the basis of the three-wire system, but the distribution circuits shall consist of two wires only except to outlets for a group of ——, —— c. p., lamps or more, and for special circuits specifically mentioned. Care shall be taken in arranging the distribution circuits to have the same number of lamps on each side of the system and that no circuit shall be connected across the outside wires. The neutral wire shall in all cases be properly tagged and shall be run between the outside wires.

106. System.

Three-Wire System Adapted to the Two-Wire System.

This specification is identical with the two-wire specification, except in "System," p. 94, sec. 84, in place of which insert the following:

107. System. All { risers, feeders } and mains shall consist of three wires, but the neutral wire shall consist of two wires each equal in cross-section to the outside wires, or of one wire equal in cross-section to the outside wires combined, in order that if desired all lights may be operated on the two-wire system; if two neutral wires are run they shall be permanently connected at each cut-out box. All distribution circuits shall consist of two wires only except to outlets for a group of ——, —— c. p., lamps or more, and for special circuits specifically mentioned. In all three-wire distribution circuits the neutral shall be equal in cross-section to the two outside wires combined. Care shall be taken in arranging the distribution circuits to have the same number of lamps on each side of the system and that no circuit shall be connected across the outside wires or between the neutral wires. The neutral shall in all cases be properly tagged and shall be run between the outside wires.

Arc System.

108. Circuits. The lights shall be divided into the following circuits:
Circuit No. 1 (Number of lights and location.)
Circuit No. 2 (Number of lights and location), etc.

109. Insulation of Wire. All wire used in the installation shall be insulated with —— for inside circuits, and with —— for outside circuits.

110. Insulation Resistance. Each circuit shall test out with an insulation resistance of at least —— ohms.

111. Joints. See p. 97, sec. 92.

112. Method of Wiring. All interior wires shall be run $\begin{Bmatrix} \text{on insulators} \\ \text{in conduits} \end{Bmatrix}$ in such a manner that the highest possible insulation is obtained. All wiring shall be neat in its mechanical

ELECTRIC LIGHTING SPECIFICATIONS.

appearance and arrangement. All exterior wires shall be run ——.

In the interior of (the) building(s) the lamps specified shall be suspended from —— securely fastened to the ceiling (and provided with a suitable device for raising and lowering). On the exterior of (the) building(s) the lamps specified shall be suspended from ——, securely attached to ——, (and provided with a suitable device for raising and lowering).

113. Suspension of Lamps.

(Add details concerning any posts, pole-steps, ornamental treatment desired, etc.)

Conduit System. Two-Wire.

114. Method of Wiring. The building shall be wired according to the system of —— and using the —— conduit manufactured by —— ——.

115. Appliances. All appliances employed shall be such as are especially adapted for use in conjunction with the conduit system.

116. Placing of Conduits. All conduits shall be placed in position { before / after } the plastering is done, and shall be firmly secured { to / within } walls and ceilings.

117. Sealing Openings. After the tubes are installed all openings in walls and floors shall be sealed so that it shall be impossible, in the event of fire, for smoke or flame to pass from one floor to another or from one room to another about the tubes.

ELECTRIC LIGHTING SPECIFICATIONS.

(See p. 94, sec. 83.)   **118. Outlets and Lights.**

(See p. 94, sec. 84.)   **119. System.**

(See p. 94, sec. 85.)   **120. Variation in Pressure.**

All single conductors shall be insulated with ——. All duplex conductors shall be insulated with ——. Duplex conductors and all single conductors larger than —— { B. & S. / B. W. G. } —— shall be stranded.   **121. Insulation.**

(See p. 95, sec. 87.)   **122. Insulation Resistance.**

From the switchboard (or centre of distribution) —— { risers / groups of risers / feeders } shall be carried to the following points: { Riser / Group / Feeder } No. 1 to ——, { riser / group / feeder } No. 2 to ——, etc.   **123. Subdivision of Risers & Feeders.**

Every conductor in each

{riser, groups of risers, feeder,} and main shall be provided with an independent tube.

{Riser Group Feeder} No. 1 shall feed all lights (location), {riser group feeder} No. 2 shall feed, etc.

124. Location of {Risers Feeders}. From the {switchboard centre of distribution} the {risers feeders} shall be carried to —— and thence upward in (channels, elevator shaft, etc., with location) to their respective junction boxes.

125. Mains (If taps terminate in secondary junction boxes). From the {riser feeder} junction boxes, mains shall be carried to secondary junction boxes where all tap lines shall centre. From {riser feeder} No. 1 shall be carried —— mains terminating at ——; from {riser feeder} No. 2, etc.

ELECTRIC LIGHTING SPECIFICATIONS.

126. Taps.

(Note.—In certain cases it is advisable to run the circuits in such a manner that no room shall be dependent on one circuit only; if so desired, it should be added under this heading.)

From the $\begin{Bmatrix} \text{riser} \\ \text{feeder} \\ \text{main} \end{Bmatrix}$ junction boxes distributing circuits shall be run to the various outlets as specified in the schedule and located on the plans. For all taps duplex conductor requiring but one tube may be employed provided the current required does not exceed —— amperes.

127. Junction Boxes.

The terminals of all the $\begin{Bmatrix} \text{risers} \\ \text{feeders} \end{Bmatrix}$ mains and taps shall be brought together in junction boxes at the points designated in this specification and on the plans [and connected with their respective cut-outs and switches; (unless the wiring contractor does not furnish cut-outs and switches, in which case add, in such a way as to permit the easy and convenient insertion of the cut-outs and switches specified.)].

128. Cut-Outs.   (See p. 99, sec. 94.)

129. Fuses.   (See p. 99, sec. 95.)

130. Switches.   (See p. 99, sec. 96.)

131. Fixture Supports.   Where no gas pipe or other support for the fixture exists, the special form of terminal box designed to furnish such support shall be employed and shall be substantially fixed to a suitable foundation in the ceiling or wall.

132. Sizes of Tubes.   All tubes shall be of sufficient size to allow the wires to be readily drawn in, withdrawn and reinserted at will.

133. Fastenings.   Tubes, whether concealed or on the surface, should be held in place by the fastenings especially designed for use with this conduit.

134. Joints.   The tubes shall be cut squarely, reamed out smoothly, and the ends joined by the use of the coupling designed for that purpose.

Where more than three elbows are unavoidable an intersection box shall be inserted to relieve both the wires and the tubes of strain when the wires are being drawn in.

135. Elbow Limitations.

All tubes shall emerge at outlets in terminal boxes leaving the outlets so protected as not to be injured by the plasterers.

136. Outlets.

To guard against mechanical injury and the destructive action of cement all floor conduits shall be made of double tube, one telescoped within the other, and both the outer and inner tubes joined in the usual manner. The outer tube shall, in the case of contact with cement, be alkali proof. As a further protection floor tubes shall be covered, during construction, with a light board. Such other precautions shall be taken to insure the safety of the tubes as the character of the building and work require.

137. Floor Work with Brass or Unarmored Conduit.

Each side of circuits carrying more than

138. Separation of Wires.

—— amperes shall be run in a separate tube. Wires forming parts of two distinct circuits shall in no case be inclosed in the same tube.

139. Meter Outlets. (See p. 100, sec. 98.)

140. Elevator Lights. (See p. 101, sec. 99.)

141. Additions and Deductions. (See p. 101, sec. 101.)

142. Instruction. (See p. 102, sec. 102.)

143. Hanging of Fixtures. (See p. 102, sec. 103.)

144. Wiring and Attaching Sockets to Fixtures Already in Place. (See p. 103, sec. 104.)

145. Suspension of Sockets and Lamps. (See p. 103, sec. 105.)

146. Three-Wire System. (See p. 105, sec. 106.)

147. Three-Wire System Adapted to the Two-Wire System. (See p. 106, sec. 107.)

ELECTRIC LIGHTING SPECIFICATIONS. 117

Interior Wiring for Central Station Plants.

This specification contemplates the complete installation of ——, —— c. p., incandescent lamps located in blocks as designated on the plans hereto attached and made a part of this specification. **148. Number of Lights.**

(See p. 91 , sec. 79.) **149. Fixtures.**

(See p. 91 , sec. 80.) **150. Lamps.**

(See p. 92 , sec. 81.) **151. Sockets.**

(Note.—Specify whether bids for two-wire direct-current systems only, for three-wire direct-current, for two-wire alternating current, or for any system will be considered.) **152. System.**

The fall of potential between the service cut-out and the most distant lamp in any building shall not exceed —— per cent. **153. Variation in Pressure.**

154. Insulation. All wires used inside of buildings shall be insulated with ——.

155. Character of the Work. All wiring shall be $\begin{Bmatrix} \text{open cleat} \\ \text{molding} \\ \text{conduit} \end{Bmatrix}$ work, neat in its mechanical appearance and arrangement.

156. Circuits. No distributing circuit shall carry more than —— amperes. In buildings requiring a greater supply of current the lights shall be divided into circuits; these circuits shall be brought together at convenient and accessible centres of distribution where all **branch cut-outs** shall be placed.

157. Cut-outs and Switches. Each branch circuit shall be provided with a double-pole cut-out. The switches specified below shall be furnished and installed All cut-outs and switches shall be mounted on incombustible bases.

(List of numbers and sizes of switches.)

158. Meters. (See p. 92, sec. 82.)

159. Suspension of Lamps.

(All) lamps shall be suspended with flexible cord pendants from double-pole ceiling cut-outs, the average length of the pendants to be ——. This contractor is to furnish all necessary —— cord, ceiling cut-outs and socket bushings. In both cut-outs and sockets the cord shall be knotted so that no weight shall come on the binding screws.

160. Molding.

(See p. 101, sec. 100.)

161. Additions and Deductions.

(See p. 101, sec. 101.)

162. Instruction.

(See p. 102, sec. 102.)

163. Hanging of Fixtures.

(See p. 102, sec. 103.)
(May require slight modification.)

164. Wiring and Attaching Sockets to Fixtures already in Place.

(See p. 103, sec. 104.)
(May require slight modification.)

POLE LINES.

Low Potential—Direct-Current System— Two or Three-Wire.

165. Franchises and Permits.
(The purchaser) shall secure all franchises, rights of way, and permits from the ——— authorities and abutting property-owners for the erection and guying of poles and stringing of wires along the routes on the map hereto attached and made a part of this specification, shall make all necessary arrangements with companies already having pole lines on any part of the same route for crossing, raising, lowering or otherwise moving their wires, and for using, moving or changing their poles, cross-arms, etc.; shall do all necessary trimming of trees; and in every reasonable way shall secure and furnish facilities for the uninterrupted continuance of the work to its completion.

The lights shall be supplied by circuits divided as follows: ———— ————, etc.

166. Division of Circuits.

All circuits shall be controlled by switches placed (location.)

167. Point of Control.

The pole line shall be composed of straight, select, shaved ———— poles, sound and free from shakes, checks or large knots; poles subject to extra strain shall be specially selected and of ample strength to bear the strain.

168. Poles.

All poles must be set ———— of their length in the ground and solidly tamped, must measure not less than ———— in diameter at the top, and the distance from the ground line to the lowest cross-arm shall be not less than ————. Corner, terminal and other poles subject to extra strain shall be securely guyed wherever possible; where impossible to guy them they shall be set with such rake and to such extra depth that the strain shall not pull them beyond the vertical position,

169. Setting and Guying.

allowance being made for the action of water and frost.

170. Distances. No two consecutive poles shall be set at a greater distance apart than —— except by special permission from ——, and all poles carrying heavy feeders or mains shall be set not more than —— apart.

171. Painting. (As desired.)

172. Gains and Cross-Arms. Gains shall be carefully cut so that the cross-arms make a snug fit and stand at right angles to the pole.

Cross-arms shall be of ——, thoroughly seasoned, sound and free from large knots; painted ——; the vertical distance between cross-arms shall not be less than ——. Double cross-arms must be placed on terminal poles and corner poles carrying wires larger than —— $\left\{ \begin{array}{l} \text{B. \& S.} \\ \text{B. W. G.} \end{array} \right\}$. At all corners making an angle greater than —— two sets of cross-arms shall be used placed at the proper angle to each other.

173. Pins and Insulators. Pins shall be of selected ——, shall fit closely in the cross-arms and be nailed in place.

Insulators shall be of glass, —— pattern, and of a size suitable for the wire they are to hold.

174. Steps. All poles on which cut-outs are placed shall be stepped.

175. Soil. (Note.—Specify character of soil, as loam, sand, etc.; also whether rock, marsh land, quicksands, etc., requiring special work.)

176. Wiring. All feeder, main and pressure wires shall be { of bare copper wire; / insulated with ——; } all service wires shall be insulated with ——.

All wires shall be so handled as to avoid kinking; wagons, drays, etc., shall not be allowed to drive over them; they shall not be dragged along the ground, over cross-arms or through trees in such a way as to injure the insulation; and shall not be allowed to sag unduly between supports, allowance being made for expansion and contraction with changes of temperature. All necessary and proper precautions shall be taken in passing over, through or near buildings of every description, through trees, crossing other lines, turning corners, etc.

Pressure wires shall be carried from the (switchboard) to each centre of distribution unless such centres are connected by an equalizing main, in which case the pressure wires shall be carried to a point on the equalizing main electrically equidistant from the centres of distribution which it connects.

177. Joints. Joints shall be mechanically strong and secure so that no movement of the two ends relatively to each other is possible, and shall be carefully sweat-soldered, the joint being wiped free from any excess of

flux; the solder shall be relied on only to give good electrical connection.

An efficient lightning arrester shall be placed on the pole, connected to the line and to a permanent ground for every —— of conductor. **178. Lightning Arresters.**

The mains shall be so proportioned that the maximum fall of potential between the centre of distribution and any service cut-out, including the loss in any transforming device, shall not exceed —— per cent. under full load. **179. Maximum Fall of Potential.**

Alternating Current System.

(Note.—The specification for the low-potential system, p. 120, et seq., may be followed in general. There should be added a specification for placing converters on poles where so required, and for running secondary mains where a single converter supplies a number of buildings, including the distance that the secondary main must be kept from the primary.) **180. Alternating System.**

Street Lighting Circuits—Arc or Incandescent.

181. Street Lighting Circuits. (Note.—The specification for the low potential system, p. 120, et seq., may be followed in general but the following additions and modifications should be introduced.)

182. Poles. (Add to sec. 168, p. 121.)

Lamp poles shall be not less than —— in length, with tops not less than —— in diameter, and set —— of their length in the ground.

183. Gains and Cross-Arms. (Add to sec. 172, p. 122.)

Where there is but a single wire on a pole a bracket may be used instead of a cross-arm. Where necessary break-arms shall be used to carry wires from the line out to the lamp.

184. Steps. (Note.—Steps may be desirable on lamp poles, if so add to sec. 174, p. 123.)

185. Suspension of Lamps. Lamps shall be suspended at the places

ELECTRIC LIGHTING SPECIFICATIONS

located on the attached map by means of (brackets, mast-arms, cross suspension, etc.) The bottom of the lamp to be not less than —— above the roadway.

The lamps and fixtures must be secured against damage or interference through ordinary wind storms, and all wires so connected that there shall be a minimum danger of short-circuiting or grounding.

The following fixtures and appliances shall be furnished and erected:

186. Fixtures (Incandescent).

—— water-proof hoods complete with reflectors, sockets, and { bracket / cross-suspension } attachments.

——goose-neck brackets, —— in length complete with post-socket or flange and the necessary guy wires.

—— sleet-proof pulleys.

—— feet —— inch weather-proof rope for raising and lowering lamps.

—— double cleats for winding up surplus rope.

———— feet ——— inch rope for suspending lamps, ———, ——— c. p. ——— volt lamps.

187. Fixtures (Arc. See also, Secs. 69, 70, 71 72, 73 and 74). ——— water-proof circuit cut-outs for cutting out circuits inside buildings from the exterior circuit. The "Off" and "On" positions shall be plainly marked and the switch shall be so constructed that when at the position marked "Off" the wires on the inside shall be entirely cut out of the main circuit.

——— outriggers, ——— in length for attaching to the exterior of buildings.

——— outriggers, ——— in length for attaching to poles.

——— { plain ornamental } pole tops.

——— mast arms, ——— in length, with proper arrangement for raising and lowering lamps. Bidders shall submit design of mast-arm recommended.

——— sleet-proof pulleys.

feet ──── inch weather-proof rope for raising and lowering lamps.

feet ── inch { weather-proof rope / stranded cable } for suspending lamps.

────── { double cleats / windlasses } for winding up surplus rope.

(Note.—The specification for wiring ^{183. Wiring.} must be made up with reference to the system or systems to be installed, whether parallel or series wiring with incandescent or arc lamps, or a combination of systems.)

(Note.—The point at which street light- ^{180. Point of Control.} ing circuits will be controlled will depend much on the system adopted and on local conditions; it is sometimes absolutely necessary that they shall be controlled from the station; in other cases for the sake of economy or convenience it is desirable to have them controlled from some other point such as the centre of distribution.)

All street lighting circuits shall be controlled by switches at (location.)

190. Additions and Deductions. While the schedule is intended to represent very closely the number of lights to be installed, yet as some changes may become necessary during the process of construction each bidder shall name in his proposal a price to be added to or deducted from the contract price for each light installed in excess of the number specified or which shall be cancelled, provided such addition or cancellation involves no change in the work already completed and shall be along the line of existing circuits. The price shall include lamp, fixtures, extra poles required and labor.

191. Instruction. (See p. 102, sec. 102.)

192. Schedules. In order to tabulate clearly and concisely the location and number of outlets for lights, switches, and meters, together with the number of lights per outlet in each individual case and the capacity of each switch and meter, also the location, catalogue number and incidental information on each fixture and shade, the attached schedules will often be found of great convenience; indeed, in making up estimates, such schedules are almost indispensable and will prove of very considerable assistance if incorporated in the specification. They will also be found useful in checking the accuracy with which the details have been taken from the plans and an aid in checking the work during construction, since they give the detailed distribution in a concise form, free from distracting explanatory clauses or directions, and are more convenient and accessible than plans; plans, too, often have the disadvantage of containing details foreign to the electrical work, which may confuse and mislead.

The shade schedule can often be incorporated in the fixture schedule. The catalogue number of fixtures is added in this schedule in order to keep in mind the exact fixture for which a given shade is intended, thus insuring against mistakes in putting them on.

In the fixture schedule, if sockets are to be furnished by another contractor, the column for same may be cancelled. As a memorandum, notes concerning the supplying of insulating joints or flanges, the wiring of fixtures, etc., may be added. The item "length" is very important and should never be omitted.

The capacity of switches is often marked in lights and is so given in the schedule; it may, however, often be advisable to designate them by their current carrying capacity in amperes to provide for the use of low volt as well as high volt lamps. The form of schedule for switches can also be used for meters, but in this case the current consumed by the lamp should be given and it should be clearly specified whether the meter is to measure direct currents only, alternating currents only, or either direct or alternating currents, also whether two or three wire.

In the lamp schedule the vertical column of outlets gives the total number for each location, the horizontal line across the bottom of the page gives the total number of outlets of each size in the building, and the sum in each case should be the same. The vertical column of lights gives the total number of lights in the building. This sum may be checked from the horizontal line of totals by multiplying those totals by the number of lights per outlet given at the top and adding the results. If the results obtained by the two methods do not agree some mistake has been made either in the arithmetical work or in placing outlets under the wrong heading; the results, to be correct, must agree. The division into "side" and "ceiling" outlets is important to the contractor, since the quantity both of labor and material required is often very largely dependent upon this relation which may also determine the method of running the circuits.

LAMP OUTLET SCHEDULE.								
	Totals.	Lights.	3	13	6	5		27
		Outlets.	2	5	3	2		12
	Light.	Ceiling.						
		Side.						
	Six Light.	Ceiling.		1				1
		Side.						
	Four Light.	Ceiling.			1			1
		Side.						
	Three Light.	Ceiling.		1		1		2
		Side.						
	Two Light.	Ceiling.	1					1
		Side.		1		1		2
	One Light.	Ceiling.			1			1
		Side.	1	2	1			4
		LOCATION.	First Floor. Hall......	Manager's Office......	Secretary's Office......	Consultation Room......	Etc......	Totals......

ELECTRIC LIGHTING SPECIFICATIONS.

SWITCH ⎱ SCHEDULE. METER ⎰					Outlets to be controlled and remarks.
			Total ⎱ Switches ⎰ ⎱ Meters. ⎰		
	No. and Capacity.				
	3 Lt.	6 Lt.	10 Lt.	Lt.	
	Location.				Totals.

FIXTURE SCHEDULE.

Location.	No. of Fixtures	Cat. No.	No. of Lights.	Length.	Finish.	Sockets Key or Keyless.	Remarks.

SHADE SCHEDULE.

Number of Shades.	Cat. No.	Cat. No. of Fixture.	No. Shade-holders.	Cat. No.	Finish.	Remarks.

STORAGE BATTERY.

This contractor shall furnish and erect, complete in all its details, the following storage battery plant.

The cells shall be of the ——— type.   193. Type.

The capacity of the battery shall be at least equal to a steady discharge at the rate of ——— amperes for ——— consecutive hours and it shall be capable of withstanding a maximum discharge at the rate of ——— amperes for ——— consecutive hours without injury.   194. Capacity.

The total number of cells shall be ———. They shall be arranged in ——— series of ——— cells each and so connected that they can be discharged at a practically uniform potential of ——— volts.   195. Number of Cells and Voltage.

(Note.—Describe method of erecting.)   196. Erection.

(Note.—Specify and describe connections for charging and discharging; switches for throwing entire battery or single cells in and out; devices for keeping the potential constant on discharging circuits; connections for testing; and any special devices and apparatuses desired.)   197. Connections.

Each bidder shall incorporate in his   198. Guarantees.

proposal such guarantees as he desires to make as to efficiency, life and cost of maintenance and repairs.

199. Summary. Each bidder shall fill out completely the following summary:

Number of cells . . . . ———
Number of series . . . . . . ———
Dimensions of each cell . . . . ———
Weight of each cell complete . ———
Rated capacity in ampere-hours . ———
Normal voltage each cell . . . ———
Normal discharging rate . . . ———
Maximum safe discharging rate . ———
Normal charging rate . . . . ———
Maximum safe charging rate . . ———

STEAM PLANT.

(Note.—This specification will be found applicable to many central station plants up to 200 h. p. or 300 h. p., and for isolated plants where the conditions imposed by municipal and insurance regulations, considerations of the utilization of exhaust steam, or of combining a heating and power plant, and questions of a similar nature do not require special attention.

The forms of Warning, Preamble, and General Specification are, with but slight changes, also adapted to this specification.)

200. *Specification for Steam Plant.*

This contractor shall furnish and erect, complete in all its details, the following steam plant:

———   { horizontal / vertical }   { single cylinder / compound }   { non-condensing / condensing }   engine(s) especially adapted for electric lighting service.

201. *Engines.*

{ This / These } engine(s) shall be capable of developing —— indicated horse-power with —— pounds initial steam pressure at the throttle, —— pounds back pressure, —— cut-off and —— revolutions per minute, and must be designed in all { their / its } parts to develop this power for indefinite periods.

The variation in speed from no load to full load shall not exceed —— per cent.

All material and workmanship shall be of the highest grade; all parts accurately made to standard gauge; all moving parts carefully balanced; all valves and packing free from leakage.

{ They / It } shall operate noiselessly and without vibration when set on (a) suitable foundation(s) and properly piped.

202. Fittings.   { Each / The } engine shall be provided with a { foundation plate, / full height foundation box, } founda-

ELECTRIC LIGHTING SPECIFICATIONS.

tion bolts and washers, sight feed cylinder lubricator, throttle valve, sight feed oil cups, full set of wrenches, governor pulley, driving pulley of proper dimensions for driving the dynamo selected (and all necessary pipes, valves, reducing motion and attachments for taking indicator cards.

Indicators to be furnished by the purchaser.)

(Note.—If the dynamo(s) $\begin{Bmatrix} are \\ is \end{Bmatrix}$ to be direct driven by the engine(s) there should be added to the above specification the type or types of dynamo(s) which may be selected from; the method of connecting engine(s) and dynamo(s); and any special fittings such as extension to engine foundation box, extended shaft, out board bearing, spot for supporting fields, attachment for securing brush-holder, etc.)

(As required.)

203. Painting.

204. Foundations. The foundation(s) for the(se) engine(s) shall be built by the {contractor / purchaser} of ———, laid ———, and shall be of sufficient length, width and depth to safely and firmly sustain {their / its} weight and all strains to which {they are / it is} subjected. All necessary excavating or filling and the removal of all debris shall be done by the {contractor. / purchaser.} The foundations shall be of such height that the driving pulley(s) will swing {clear of / below} the floor ———.

(Note.—Foundations for direct-connected outfits should extend under both engine and dynamo.)

(Character of the soil should also be specified, and the method of isolating and insulating the foundation if required.)

ELECTRIC LIGHTING SPECIFICATIONS. 143

205. Starting Plant and Instruction.

(Note.—If this contractor furnishes plant complete, this section can be put before final summary.)

When the $\begin{Bmatrix} \text{engine(s)} \\ \text{plant} \end{Bmatrix}$ $\begin{Bmatrix} \text{are} \\ \text{is} \end{Bmatrix}$ ready for operation $\begin{Bmatrix} \text{they} \\ \text{it} \end{Bmatrix}$ shall be run for a period of —— days by competent engineers furnished by the contractor. $\begin{Bmatrix} \text{Assistant Engineer,} \\ \text{Fireman,} \end{Bmatrix}$ oil, waste, etc., will be furnished by (the purchaser).

This contractor shall also give all necessary instructions to the engineer of (the purchaser) for the proper care, maintenance, and operation of the $\begin{Bmatrix} \text{engine(s),} \\ \text{plant,} \end{Bmatrix}$ such instructions to be given during the trial period stipulated above.

206 Belts.

—— $\begin{Bmatrix} \text{double leather, endless, solid} \\ \text{\quad''\qquad''\qquad''\qquad perforated} \\ \text{link} \\ \text{rubber} \\ \text{cotton-leather} \\ \text{rope, etc.,} \end{Bmatrix}$ belts free from defects of any kind, —— in width by —— in length, and capable of transmitting —— horse-power at a belt speed of ——.

207. Renewal Parts. Such renewal parts as it is advisable to keep on hand, adding hereto an itemized list of same.

208. Summary. (See Note: p. 52, sec. 27.)

Each bidder shall fill out completely the following summary:

Number of engines . . . . . . ——
Size of cylinder(s) . . . . . ——
Diameter of steam pipe . . . ——
Diameter of exhaust pipe . . . ——
Floor space . . . . ——
Indicated horse-power at —— revolutions, —— initial steam pressure, —— back pressure, cutting off at —— stroke . . . ——
Speed . . . . ——
Maximum variation in speed between no load and full load——
Number of pulleys . . . . . ——

ELECTRIC LIGHTING SPECIFICATIONS.

Dimensions of pulleys . . . . ——

209. Counter-Shafting.

——of $\begin{cases} \text{turned steel,} \\ \text{hammered iron,} \\ \text{etc.,} \end{cases}$ shafting ——

in diameter.

—— $\begin{cases} \text{floor stands,} \\ \text{drop hangers,} \\ \text{post hangers,} \\ \text{pedestals, etc.,} \end{cases}$ complete with

adjustable, (self-oiling) boxes, base-plates, bolts, etc. $\begin{cases} \text{Height} \\ \text{Drop} \end{cases}$ from $\begin{cases} \text{floor} \\ \text{ceiling} \end{cases}$ to centre line of shaft ——.

Shafting to be key-seated for the pulleys specified below, and provided with all necessary collars, guard rings, etc.

210. Pulleys.

—— $\begin{cases} \text{plain cast iron,} \\ \text{split "   "} \\ \text{plain wood,} \\ \text{split   "} \\ \text{grooved, etc.} \end{cases}$ pulleys accur-

ately bored, turned, balanced, and provided with key seats and keys. Pulleys to be —— in diameter by —— face, and capable of transmitting —— horse-power at —— revolutions.

——, —— arm, balanced friction-clutch pulleys, —— in diameter by —— face, and capable of transmitting —— horsepower at —— revolutions. Each pulley to be provided with a —— shifter rig, to operate from ——. The clutch must pick up the load without shock or jar, and the shifter rig must be positive in its action, not liable to get out of order, free from any tendency or liability to be thrown in or out accidentally.

(—— friction-clutch couplings, cut-off couplings, compression couplings, plate couplings, jaw clutches, etc., also idlers and method of applying.)

211. Foundations. (Note.—Specify character of foundation upon which shafting is to be placed, whether special foundations of brick or stone, floor timbers, walls, ceilings, posts, etc.)

212. Boiler (s). ——, ——, —— boiler(s) (each) rated at —— horse-power.

The boiler(s) shall contain not less than —— of heating surface per rated horse power. The shell(s) shall be made of steel having a tensile strength of not less than ——, and shall be braced and stayed for a working pressure of —— pounds, and tested to —— pounds hydrostatic pressure.

{ They / It } shall be inspected and insured by a responsible steam boiler insurance company, this contractor to furnish a certificate of inspection and a policy of insurance for ——.

{ Each / The } boiler shall be provided with {dome or drums / dry pipe} the necessary lugs, brackets, plates, bolts, stays, anchor and binder rods, man and hand holes, steam and blow-off connections, stop and safety valves, pressure gauge, water column, gauge cocks and drip, set of stoking tools consisting of ——, and all other appliances properly coming under this head. ^{213. Fittings}

The boiler(s) shall be set in masonry in a first-class manner, according to standard ^{214. Setting.}

practice. This contractor to do all necessary excavating or filling for the foundation(s) and to remove all debris.

(Specify character of soil upon which boilers will be placed.)

215. Stack. (Note.—Specify material, diameter, height, lining, base, location, etc.)

216. Smoke Connections. Smoke connections of ——, having an area ——, are to be made with the stack, and are to be provided with —— dampers and —— cleaning doors arranged as follows : ——

217. Piping. (Note.—If the piping is to be erected with reference to existing piping or to meet special conditions the specification must be made up with due regard to such conditions; for entirely new piping, independent of auxiliary connections, the following specification will cover many cases.)

This contractor shall furnish and erect

all necessary and proper piping to pipe the entire plant. From each boiler a branch pipe shall be carried to a main header located ——, and from this header a branch pipe shall be carried to each engine; all branch pipes from boilers and to engines shall be provided with a —— valve so that any boiler or any engine may be cut out without interfering with the operation of the remainder. The valve in the engine branch shall be in addition to the throttle at the engine. The engine exhausts shall be connected to an exhaust header and each exhaust branch shall be provided with a —— valve; the exhaust header shall be connected with the heater and from the heater shall be carried ——. (For non-condensing engines carry to atmosphere; for condensing engines to the condenser.) The exhaust from the (condenser and) feed pumps shall be carried ——. (For non-condensing engines carry to exhaust header or atmosphere; for condensing engines carry to the condenser or to an auxiliary heater.) Feed (and condensing) water will be brought to —— by the purchaser). Drip (and condenser discharge) connections shall be made at ——. All piping —— and larger shall be flanged. (Add specification as to hot-well, exhaust head, back-pressure valve, automatic atmospheric relief valve, connections for auxiliary heater, etc., if required; also the disposal of drips from live steam pipes, separator, engine jackets, receiver, etc.)

All live steam piping —— and larger, (and also the following sections of exhaust piping —— ——) shall be neatly covered with a substantial, non-heat-conducting pipe covering.

218 Condensers. —— { independent steam driven / —————— driven }
{ jet / surface } condenser(s) complete with —— pump(s) and all connections; { each / the } condenser to have a capacity of —— pounds of steam condensed per hour

ELECTRIC LIGHTING SPECIFICATIONS.

with condensing water at —— degrees
{ Fahr. }
{ Cent. }

——, —— feed pump(s) of standard manufacture (each) having ample capacity to feed — h. p. in boilers. 219. Feed Pump(s) and Injector(s).

—— feed-water heater(s) [and purifier(s)] of standard manufacture (each) having ample capacity for —h. p. in boilers. 220. Feed-Water Heater (s) and Purifier(s).

(Specify character of water to be used, and whether exhaust steam is to be used for any other purpose than heating feed-water.)

—— separator(s) of standard manufacture. 221. Separator(s).

A gauge board of —— shall be erected in ——, and the following instruments mounted thereon: ——. Connections shall be made between these instruments and —— in the following manner ——. 222. Gauge Board.

223. Painting. (As required.)

224. Renewal Parts. Such renewal parts as it is advisable to have on hand shall be furnished, a list of same to be added hereto.

225. Summary. (See Note: p. 52, sec. 27.)

Each bidder shall fill out completely the following summary:

Number of boilers . . . ———
Rated horse-power . . . — —
Kind . . . . . . ———
Diameter of shell . . . ———
Length of shell . . . . ———
Number of tubes . . . ———
Diameter of tubes . . . ———
Heating surface . . . ———
Tensile strength of steel . . ———
Working pressure . . . ———
Testing pressure . . . ———
Outside dimensions of setting . ———
Number of condensers . . ———
Kind . . . . . ———
Capacity of each . . . ———
Number of pumps . . . ———

Kind . . . . . ———
Capacity of each . . . ———
Number of heaters . . . ———
Kind . . . . . ———
Capacity of each . . . ———

RULES AND REQUIREMENTS

OF THE
NATIONAL BOARD OF FIRE UNDERWRITERS FOR THE INSTALLATION OF WIRING AND APPARATUS FOR ELECTRIC LIGHT AND POWER AS RECOMMENDED BY THE UNDERWRITERS' NATIONAL ELECTRIC ASSOCIATION.

EDITION OF JAN. 1, 1895.

The use of wire ways for rendering concealed wiring permanently accessible, is most heartily endorsed and recommended; and this method of accessible concealed construction is advised for general use.

Architects are urged, when drawing plans and specifications, to make provision for the channeling and pocketing of buildings for electric light or power wires, and in specifications for electric gas lighting to require a two-wire circuit, whether the building is to be wired for electric lighting or not, so that no part of the gas fixtures or gas piping be allowed to be used for the gas lighting circuit.

CENTRAL STATIONS.
Class A.
FOR LIGHT OR POWER.

These Rules also apply to Dynamo Rooms in Isolated Plants, connected with or detached from buildings used for other purposes; also to all varieties of apparatus therein of both high and low potential.

1. GENERATORS:—

   a. Must be located in a dry place.

   b. Must be insulated on floors or base-frames, which must be kept filled, to prevent absorption of moisture, and also kept clean and dry.

   c. Must never be placed in a room where any hazardous process is carried on, more in places where they would be exposed to inflammable gases, or flyings, or combustible material.

   d. Must each be provided with a waterproof covering.

2. CARE AND ATTENDANCE:—

   A competent man must be kept on

duty in the room where generators are operating.

Oily waste must be kept in *approved* metal cans, and removed daily. (See Definitions).

3. CONDUCTORS:—

From generators, switchboards, rheostats, or other instruments, and thence to outside lines, conductors—

a. Must be in plain sight, and readily accessible.

b. Must be wholly on non-combustible insulators, such as glass or porcelain.

c. Must be separated from contact with floors, partitions or walls, through which they may pass, by non-combustible insulating tubes, such as glass or porcelain.

d. Must be kept rigidly so far apart that they cannot come in contact.

e. Must be covered with non-inflammable insulating material sufficient to prevent accidental contact, except that " bus bars " may be made of bare metal.

f. Must have ample carrying capacity, to prevent heating. (See Capacity of Wires Table.)

4. SWITCHBOARDS:—

a. Must be so placed as to reduce to a minimum the danger of communicating fire to adjacent combustible material. (See Definitions).

b. Must be accessible from all sides when the connections are on the back; or may be placed against a brick or stone wall when the wiring is entirely on the face.

c. Must be kept free from moisture.

d. Must be made of non-combustible material, or of hard wood in skeleton form, filled to prevent absorption of moisture.

e. Bus bars must be equipped in accordance with Rule 3 for placing conductors.

5. RESISTANCE BOXES AND EQUALIZERS:—

a. Must be equipped with metal, or other non-combustible frames. (See Definitions).

b. Must be placed on the switchboard, or, if not thereon, at a distance of a foot from combustible material, or separated therefrom by a non-inflammable, non-absorptive, insulating material.

6. LIGHTNING ARRESTERS:—

a. Must be attached to each side of every overhead circuit connected with the station.

b. Must be mounted on non-combustible bases in plain sight on the switchboard, or in an equally accessible place, away from combustible material.

c. Must be connected with at least two " earths" by separate wires, not smaller than No. 6 B. & S., which must not be connected to any pipe within the building, and must be run as nearly as possible in a straight line from the arresters to the earth connection.

d. Must be so constructed as not to maintain an arc after the discharge has passed.

7. TESTING:—

a. All series and alternating circuits

must be tested every two hours while in operation, to discover any leakage to earth, abnormal in view of the potential and method of operation.

b. All multiple arc low potential systems (300 volts or less) must be provided with an indicating or detecting device, readily attachable, to afford easy means of testing where the station operates continuously.

c. Data obtained from all tests must be preserved for examination by insurance inspectors.

These rules on testing to be applied at such places as may be designated by the association having jurisdiction.

MOTORS.

8. MOTORS:—

a. Must be wired under the same precautions as with a current of the same volume and potential for lighting. The motor and resistance box must be protected by a double pole cut-out and controlled by a double pole switch, except in cases where one-quarter horse-power or less is used on low tension circuit, a single pole switch will be accepted.

b. Must be thoroughly insulated, mounted on filled dry wood, be raised at least eight inches above the surrounding floor, be provided with pans to prevent oil from soaking into the floor, and must be kept clean.

c. Must be covered with a waterproof cover when not in use, and, if deemed necessary by the Inspector, be enclosed in an *approved* case. (See Definitions).

9. RESISTANCE BOXES:—

a. Must be equipped with metal or other non-combustible frames. (See Definitions).

b. Must be placed on the switchboard, or at a distance of a foot from combustible material, or separated therefrom by a non-inflammable, non-absorptive, insulating material.

Class B.
HIGH POTENTIAL SYSTEMS.
OVER 300 VOLTS.

Any circuit attached to any machine, or combination of machines, which develop over 300 volts difference of potential between any two wires, shall be considered as a high potential circuit and coming under that class, unless an approved transforming device is used, which cuts the difference of potential down to less than 300 volts.

10. OUTSIDE CONDUCTORS. All outside overhead conductors (including services):—

a. Must be covered with some *approved* insulating material, not easily abraded, firmly secured to properly insulated and substantially built supports, all tie wires having an insulation equal to that of the conductors they confine. (See Definitions).

b. Must be so placed that moisture cannot form a cross connection between them, not less than a foot apart, and not in contact with any substance other than their insulating supports.

c. Must be at least seven feet above the highest point of flat roofs, and at least one foot above the ridge of pitched roofs over which they pass or to which they are attached.

d. Must be protected by *dead insulated guard irons* or *wires* from possibility of contact with other conducting wires or substances to which current may leak. Special precautions of this kind must be taken where sharp angles occur, or where any wires might possibly come in contact with electric light or power wires.

e. Must be provided with petticoat insulators of glass or porcelain. Porcelain knobs or cleats and rubber hooks will not be approved.

f. Must be so spliced or joined as to be both mechanically and electrically secure without solder. The joints must then be soldered, to insure preservation, and covered with an insulation equal to that on the conductors. (See Definitions).

g. Telegraph, telephone, and similar wires must not be placed on the same cross-arm with electric light or power wires.

11. SERVICE BLOCKS:—

Must be covered over their entire surface with at least two coats of waterproof paint.

INTERIOR CONDUCTORS.

12. ALL INTERIOR CONDUCTORS:—

a. Must be covered where they enter buildings from outside terminal insulators to and through the walls, with extra waterproof insulation, and must have drip loops outside. The hole through which the conductor passes must be bushed with waterproof and non-combustible insulating tube, slanting upward toward the inside. The tube must be sealed with tape, thoroughly painted, and securing the tube to the wire.

b. Must be arranged to enter and leave the building through a double contact service switch, which will effectually close the main circuit and disconnect the interior wires when it is turned "off." The switch must be so constructed that it shall be automatic in its action, not stopping between points when started, and prevent an arc between the points under all circumstances; it must indicate on inspection whether the current be "on" or "off," and be mounted in a non-combustible case, and kept free from moisture, and easy of access to police or firemen. So-called "snap switches" shall not be used on high potential circuits.

c. Must be always in plain sight, and never encased, except when *required* by the Inspector.

d. Must be covered in all cases with an *approved* non-combustible material that will adhere to the wire, not fray by friction, and bear a temperature of 150° F. without softening. (See Definitions).

e. Must be supported on glass or porcelain insulators, and kept rigidly at least eight inches from each other, except within the structure of lamps or on hanger boards, cut-out boxes, or the like, where less distance is necessary.

f. Must be separated from contact

with walls, floors, timbers, or partitions through which they may pass by non-combustible insulating tube.

g. Must be so spliced or joined as to be both mechanically and electrically secure without solder. They must then be soldered, to insure preservation, and covered with an insulation equal to that on the conductors.

LAMPS AND OTHER DEVICES.

13. ARC LAMPS—In every case:—

a. Must be carefully isolated from inflammable material.

b. Must be provided at all times with a glass globe surrounding the arc, securely fastened upon a closed base. No broken or cracked globes to be used.

c. Must be provided with an *approved* hand switch, also an automatic switch, that will shunt the current around the carbons should they fail to feed properly. (See Definitions).

d. Must be provided with reliable stops to prevent carbons from falling out in case the clamps become loose.

e. Must be carefully insulated from the circuit in all their exposed parts.

f. Must be provided with a wire netting around the globe, and an *approved* spark arrester above to prevent escape of sparks, melted copper, or carbon, where readily inflammable material is in the vicinity of the lamps. It is recommended that plain carbons, not copper plated, be used for lamps in such places. (See Definitions).

g. Hanger boards must be so constructed that all wires and current-carrying devices thereon shall be exposed to view, and thoroughly insulated by being mounted on a waterproof, non-combustible substance. All switches attached to the same must be so constructed that they shall be automatic in their action, not stopping between points when started, and preventing an arc between points under all circumstances.

h. Where hanger boards are not used, lamps to be hung from insulated supports other than their conductors.

14. INCANDESCENT LAMPS IN SERIES CIRCUITS HAVING A MAXIMUM POTENTIAL OF 300 VOLTS OR OVER:—

a. Must be governed by the same rules as for arc lights, and each series lamp provided with an *approved* hand-spring switch and automatic cut-out.

b. Must have each lamp suspended from a hanger board by means of a rigid tube.

c. No electromagnetic device for switches and no system of multiple series or series multiple lighting will be approved.

d. Under no circumstances can series lamps be attached to gas fixtures.

Class C.
LOW POTENTIAL SYSTEMS.
300 VOLTS OR LESS.
OUTSIDE CONDUCTORS.

15. OUTSIDE OVERHEAD CONDUCTORS:—

a. Must be erected in accordance with the rules for (high potential) conductors.

b. Must be separated not less than 12 inches, and be provided with an *approved* fusible cut-out, that will cut off the entire current as near as possible to the entrance to the building and inside the walls. (See Definitions).

16. UNDERGROUND CONDUCTORS:—

a. Must be protected against moisture and mechanical injury, and be removed at least two feet from combustible material when brought into a building, but not connected with the interior conductors.

b. Must have a switch and a cut-out for each wire between the underground conductors and the interior wiring when the two parts of the wiring are connected.

These switches and fuses must be placed as near as possible to the end of the underground conduit, and connected therewith by specially insulated conductors, kept apart not less than two and a half inches. (See Definitions).

c. Must not be so arranged as to shunt the current through a building around any catch-box.

INSIDE WIRING.

GENERAL RULES.

17. At the entrance of every building there shall be an *approved* switch placed in the service conductors by which the current may be entirely cut off. (See Definitions).

18. CONDUCTORS:—

a. Must have an *approved* insulating covering, and must not be of sizes smaller than No. 14 B. & S., No. 16 B. W. G., or No. 4 E. S. G., except that in conduit installed under Rule 22, No. 16 B. & S., No. 18 B. W. G., or No. 4 E. S. G., may be used. (See Definitions).

b. Must be protected when passing through floors; or through walls, partitions, timbers, etc., in places liable to be exposed to dampness by waterproof, noncombustible, insulating tubes, such as glass or porcelain.

Must be protected when passing through walls, partitions, timbers, etc., in places not liable to be exposed to dampness by *approved* insulating bushings specially made for the purpose.

c. Must be kept free from contact with gas, water, or other metallic piping, or any other conductors or conducting material which they may cross (except high potential conductors) by some continuous and firmly fixed non-conductor creating a separation of at least one inch. Deviations from this rule may sometimes be allowed by special permission.

d. Must be so placed in crossing high potential conductors that there shall be a space of at least one foot at all points between the high and low tension conductors.

e. Must be so placed in wet places that an air space will be left between conductors and pipes in crossing, and the former must be run in such a way that they cannot come in contact with the pipe accidentally. Wires should be run *over* all pipes upon which condensed moisture is likely to gather, or which by leaking might cause trouble on a circuit.

ELECTRIC LIGHTING SPECIFICATIONS.

f. Must be so spliced or joined as to be both mechanically and electrically secure without solder. They must then be soldered, to insure preservation, and covered with an insulation equal to that on the conductors. (See Definitions).

SPECIAL RULES.

19. WIRING NOT ENCASED IN MOLDING OR APPROVED CONDUIT:—

a. Must be supported wholly on non-combustible insulators, constructed so as to prevent the insulating coverings of the wire from coming in contact with other substances than the insulating supports.

b. Must be so arranged that wires of opposite polarity, with a difference of potential of 150 volts or less, will be kept apart at least two and one-half inches.

c. Must have the above distance increased proportionately where a higher voltage is used.

d. Must not be laid in plaster, cement or similar finish.

e. Must never be fastened with staples.

IN UNFINISHED LOFTS, BETWEEN FLOORS AND CEILINGS, IN PARTITIONS, AND OTHER CONCEALED PLACES.

f. Must have at least one inch clear air space surrounding them.

g. Must be at least ten inches apart when possible, and should be run singly on separate timbers or studding.

h. Wires run as above immediately under roofs, in proximity to water tanks or pipes, will be considered as exposed to moisture.

i. When from the nature of the case it is impossible to place concealed wire on non-combustible insulating supports of glass or porcelain, the wires may be fished on the loop system, if encased throughout in *approved* continuous flexible tubing or conduit.

j. Wires must not be fished for any great distance, and only in places where the inspector can satisfy himself that the above rules have been complied with.

k. Twin wires must never be employed in this class of concealed work.

20. MOLDINGS:—

a. Must never be used in concealed work or in damp places.

b. Must have at least two coats of waterproof paint or be impregnated with a moisture repellant.

c. Must be made of two pieces, a backing and capping, so constructed as to thoroughly encase the wire, and maintain a distance of one-half inch between conductors of opposite polarity, and afford suitable protection from abrasion.

21. SPECIAL WIRING:—

In breweries, packing houses, stables, dye-houses, paper and pulp mills, or other buildings specially liable to moisture or acid, or other fumes liable to injure the wires or insulation, except where used for pendants, conductors—

a. Must be separated at least six inches.

b. Must be provided with an *approved* waterproof covering. (See Definitions).

c. Must be carefully put up.

d. Must be supported by glass or porcelain insulators. No switches or fusible cut-outs will be allowed where exposed to inflammable gases or dust, or to flyings of combustible material.

e. Must be protected when passing through floors, walls, partitions, timbers, etc., by waterproof, non-combustible, insulating tubes, such as glass or porcelain.

22. INTERIOR CONDUITS*:—(See · Definitions).

a. Must be continuous from one junction box to another, or to fixtures, and must be of material that will resist the fusion of the wire or wires they contain without igniting the conduit.

b. Must not be of such material or construction that the insulation of the conductor will ultimately be injured or destroyed by the elements of the composition.

---

* The object of a tube or conduit is to facilitate the insertion or extraction of the conductors to protect them from mechanical injury, and, as far as possible, from moisture. Tubes or conduits are to be considered merely as raceways, and are not to be relied on for insulation between wire and wire, or between the wire and the ground.

c. Must be first installed as a complete conduit system, without conductors, strings, or anything for the purpose of drawing in the conductors, and the conductors then to be pushed or fished in. The conductors must not be placed in position until all mechanical work on the building has been, as far as possible, completed.

d. Must not be so placed as to to be subject to mechanical injury by saws, chisels, or nails.

e. Must not be supplied with a twin conductor, or two separate conductors, in a single tube. (See page 198, Rule 22).

f. Must have all ends closed with good adhesive material, either at junction boxes or elsewhere, whether such ends are concealed or exposed. Joints must be made air-tight and moisture proof.

g. Conduits must extend at least one inch beyond the finished surface of walls or ceilings until the mortar or other similar material be entirely dry, when the projection may be reduced to half an inch.

23. DOUBLE POLE SAFETY CUT-OUTS:—

a. Must be in plain sight or enclosed in an *approved* box readily accessible. (See Definitions).

b. Must be placed at every point where a change is made in the size of the wire (unless the cut-out in the larger wire will protect the smaller).

c. Must be supported on bases of non-combustible, insulating, moisture-proof material.

d. Must be supplied with a plug (or other device for enclosing the fusible strip or wire) made of non-combustible and moisture-proof material, and so constructed that an arc cannot be maintained across its terminals by the fusing of the metal.

e. Must be so placed that on any combination fixture no group of lamps requiring a current of six amperes or more shall be ultimately dependent upon one cut-out. Special permission may be given *in writing* by the Inspector for departure from this rule in case of large chandeliers.

f. All cut-out blocks must be stamped with their *maximum* safe-carrying capacity in amperes.

24. SAFETY FUSES:—

a. Must all be stamped or otherwise marked with the number of amperes they will carry indefinitely without melting.

b. Must have fusible wires or strips (where the plug or equivalent device is not used), with contact surfaces or tips of harder metal, soldered or otherwise, having perfect electrical connection with the fusible part of the strip.

c. Must all be so proportioned to the conductors they are intended to protect that they will melt before the maximum safe-carrying capacity of the wire is exceeded.

25. TABLE OF CAPACITY OF WIRES:—

It must be clearly understood that the size of the fuse depends upon the size of the smallest conductor it protects, and not upon the amount of current to be used on the circuit. Below is a table showing the safe-carrying capacity of conductors of different sizes in Brown & Sharpe gauge, which must be followed in the placing of interior conductors:

	Table A. Concealed Work.	Table B. Open Work.
B. & S.	Amperes.	Amperes.
0000	218	312
000	181	262
00	150	220
0	125	185
1	105	156
2	88	131
3	75	110
4	63	92
5	53	77
6	45	65
8	33	46
10	25	32
12	17	23
14	12	16
16	6	8
18	3	5

NOTE.—By "open work" is meant construction which admits of all parts of the surface of the insulating covering of the wire being surrounded by *free* air. The carrying capacity of 16 and 18 wire is given, but no wire smaller than 14 is to be used except as allowed under Rules 18 (*a*) and 27 (*d*).

26. SWITCHES:—

a. Must be mounted on moisture-proof and non-combustible bases, such as slate or porcelain.

b. Must be double pole when the circuits which they control supply more than six 16-candle-power lamps, or their equivalent.

c. Must have a firm and secure contact; must make and break readily, and not stop when motion has once been imparted by the handle.

d. Must have carrying capacity sufficient to prevent heating.

e. Must be placed in dry, accessible places, and be grouped as far as possible, being mounted—when practicable—upon slate or equally non-combustible backboards. Jackknife switches, whether provided with friction or spring stops, must be so placed that gravity will tend to open rather than close the switch.

FIXTURE WORK.

27. *a.* In all cases where conductors are concealed within or attached to gas fixtures, the latter must be insulated from

the gas-pipe system of the building by means of *approved* joints. The insulating material used in such joints must be of a substance not affected by gas, and that will not shrink or crack by variation in temperature. Insulating joints, with soft rubber in their construction, will not be approved. (See Definitions).

b. Supply conductors, and especially the splices to fixture wires, must be kept clear of the grounded part of gas pipes, and where shells are used the latter must be constructed in a manner affording sufficient area to allow this requirement.

c. When fixtures are wired outside, the conductors must be so secured as not to be cut or abraded by the pressure of the fastenings or motion of the fixture.

d. All conductors for fixture work must have a waterproof insulation that is durable and not easily abraded, and must not in any case be smaller than No. 18 B. & S., No. 20 B. W. G., No. 2 E. S. G.

e. All burrs or fins must be removed

before the conductors are drawn into a fixture.

f. The tendency to condensation within the pipes should be guarded against by sealing the upper end of the fixture.

g. No combination fixture in which the conductors are concealed in a space less than one-fourth inch between the inside pipe and the outside casing, will be approved.

h. Each fixture must be tested for "contacts" between conductors and fixtures, for "short circuits," and for ground connections before the fixture is connected to its supply conductors.

i. Ceiling blocks of fixtures should be made of insulating material; if not, the wires in passing through the plate must be surrounded with hard rubber tubing.

28. ARC LIGHTS ON LOW POTENTIAL CIRCUITS:—

a. Must be supplied by branch conductors not smaller than No. 12 B. & S. gauge.

b. Must be connected with main conductors only through double pole cut-outs.

c. Must only be furnished with such resistances or regulators as are enclosed in non-combustible material, such resistances being treated as stoves.

Incandescent lamps must not be used for resistance devices.

d. Must be supplied with globes and protected as in the case of arc lights on high potential circuits.

29. ELECTRIC GAS LIGHTING:—

Where electric gas lighting is to be used on the same fixture with the electric light—

a. No part of the gas piping or fixture shall be in electrical connection with the gas lighting circuit.

b. The wires used with the fixtures must have a non-inflammable insulation, or, where concealed between the pipe and shell of the fixture, the insulation must be such as required for fixture wiring for the electric light.

c. The whole installation **must test** free from " grounds."

d. **The** two installations **must** test perfectly free from connection with each other.

30. SOCKETS:—

a. No portion of the lamp socket exposed to contact with outside objects must be allowed to come into electrical contact with either of the conductors.

b. In rooms where inflammable gases may exist, or where the atmosphere is damp, the incandescent lamp and socket should be enclosed in a vapor-tight globe.

31. FLEXIBLE CORD:—

a. Must be made of conductors, each surrounded with a moisture-proof and a non-inflammable layer, and further insulated from each other by a mechanical separator of carbonized material. Each of these conductors must be composed of several strands.

b. Must not sustain more than one

light not exceeding 50 candle-power.

c. Must not be used except for pendants, wiring of fixtures, and portable lamps or motors.

d. Must not be used in show windows.

e. Must be protected by insulating bushings where the cord enters the socket. The ends of the cord must be taped, to prevent fraying of the covering.

f. Must be so suspended that the entire weight of the socket and lamp will be borne by knots under the bushing in the socket, and above the point where the cord comes through the ceiling block or rosette, in order that the strain may be taken from the joints and binding screws.

g. Must be equipped with keyless sockets as far as practicable, and be controlled by wall switches.

RULE 32. DECORATIVE SERIES LAMPS.

Incandescent lamps run in series circuits shall not be used for decorative purposes inside of buildings.

Class D.

ALTERNATING SYSTEMS. — CONVERTERS OR TRANSFORMERS.

33. CONVERTERS:—

a. Must not be placed inside of any building, except the Central Station, unless by special permission of the Underwriters having jurisdiction.

b. Must not be placed in any but metallic or other non-combustible cases.

c. Must not be attached to the outside walls of buildings, unless separated therefrom by substantial insulating supports. IN THOSE CASES WHERE IT MAY NOT BE POSSIBLE TO EXCLUDE THE CONVERTERS AND PRIMARY WIRES ENTIRELY FROM THE BUILDING, THE FOLLOWING PRECAUTIONS MUST BE STRICTLY OBSERVED:—

34. Converters must be located at a point as near as possible to that at which the primary wires enter the building, and must be placed in a room or vault constructed of, or lined with, fire-resisting material, and used only for the purpose. They must be effectually insulated from

the ground, and the room in which they are placed be practically air-tight, except that it shall be thoroughly ventilated to the out-door air, if possible, through a chimney or flue.

35. PRIMARY CONDUCTORS:—

a. Must each be heavily insulated with a coating of moisture-proof material from the point of entrance to the transformer, and, in addition, must be so covered and protected that mechanical injury to them, or contact with them, shall be practically impossible.

b. Must each be furnished, if within a building, with a switch and a fusible cut-out where the wires enter the building, or where they leave the main line, on the pole or in the conduit. These switches should be enclosed in secure and fire-proof boxes, preferably outside the building.

c. Must be kept apart at least ten inches, and at the same distance from all other conducting bodies when inside a building.

36. Secondary Conductors:—

Must be installed according to the rules for "Low Potential Systems."

Class E.
ELECTRIC RAILWAYS.

37. All rules pertaining to arc-light wires and stations shall apply (so far as possible) to street railway power stations and their conductors in connection with them.

38. Power Stations:—

Must be equipped in each circuit as it leaves the station with an *approved* automatic "breaker," or other device that will immediately cut off the current in case the trolley wires become grounded. This device must be mounted on a fire-proof base and in full view and reach of the attendant. (See Definitions).

39. Trolley Wires:—

a. Must be no smaller than No. 0 B. & S. copper, or No. 4 B. & S. silicon bronze, and must readily stand the strain put upon them when in use.

b. Must be well insulated from their supports, and in case of the side or double pole construction, the supports shall also be insulated from the poles immediately outside of the trolley wire.

c. Must be capable of being disconnected at the power house, or of being divided into sections, so that in case of fire on the railway route the current may be shut off from the particular section, and not interfere with the work of the firemen. This rule also applies to *feeders*.

d. Must be safely protected against contact with all other conductors.

40. CAR WIRING:—

Must be always run out of reach of the passengers, and must be insulated with a waterproof insulation.

41. LIGHTING AND POWER FROM RAILWAY WIRES:—

Must not be permitted, under any pretense, in the same circuit with trolley wires with a ground return, nor shall the same dynamo be used for both purposes, except in street railway cars, electric car houses, and their power stations.

42. Car Houses:—

Must have special cut-outs located at a proper distance outside, so that all circuits within any car house can be cut out at one point.

43. Ground Return Wires:—

Where ground return is used it must be so arranged that no difference of potential will exist greater than 5 volts to 50 feet, or 50 volts to the mile between any two points in the earth or pipes therein.

Class F.

44. Storage or Primary Batteries:—

a. When current for light and power is taken from primary or secondary batteries, the same general regulations must be observed as apply to similar apparatus fed from dynamo generators developing the same difference of potential.

b. All secondary batteries must be mounted on *approved* insulators.

c. Special attention is directed to the rules (page 173) for rooms where acid fumes exist.

d. The use of any metal liable to corrosion must be avoided in connections of secondary batteries.

MISCELLANEOUS.

45. *a.* The wiring in any building must test free from grounds; *i. e.*, each main supply line and every branch circuit shall have an insulation resistance of at least 25,000 ohms, and should have an insulation resistance between conductors and between all conductors and the ground (not including attachments, sockets, receptacles, etc.), of not less than the following:—

Up to 10 amperes	. . . .	4,000,000
" 25 "	. . . .	1,600,000
" 50 "	. . . .	800,000
" 100 "	. . . .	300,000
" 200 "	. . . .	160,000
" 400 "	. . . .	80,000
" 800 "	. . . .	22,000
" 1600 "	. . . .	11,000

All cut-outs and safety devices in place in the above.

Where lamp sockets, receptacles and electroliers, etc., are connected, one-half of the above will be required.

b. Ground wires for lightning arresters of all classes, and ground detectors, must not be attached to gas pipes within the building.

c. Where telephone, telegraph, or other wires connected with outside circuits are bunched together within any building, or where inside wires are laid in conduit or duct with electric light or power wires, the covering of such wires must be fire-resisting, or else the wires must be enclosed in an air-tight tube or duct.

d. All conductors connecting with telephone, district messenger, burglar alarm, watch clock, electric time, and other similar instruments, must be provided near the point of entrance to the building with some protective device which will operate to shunt the instruments in case of a dangerous rise of potential, and will open the circuit and arrest an abnormal current flow. Any conductor normally forming an innocuous circuit may become a source of fire hazard if crossed with another conductor, through which it may become charged with a relatively high pressure. (See Definitions).

e. The following formula for soldering fluid is suggested:—

    Saturated solution of zinc.............. 5 parts
    Alcohol........................... 4 parts
    Glycerine .... 1 part

DEFINITIONS.

Definitions of the word APPROVED as used in these Rules, and notice of the approval of certain wires and materials, and the interpretation of certain rules.

RULE 2.   CARE AND ATTENDANCE:—

Approved waste cans shall be made of metal, with legs raising can three inches from the floor and with self-closing covers.

RULE 4.   SWITCHBOARDS:—

Section *a*. Special attention is called to the fact that switchboards should not be built down to the floor, nor up to the ceiling, but a space of at least eighteen inches, or two feet, should be left between the floor and the board, and between the ceiling and the board, in order to prevent fire from communicating from the switchboard to the floor or ceiling, and also to prevent the forming of a partially concealed space very liable to be used for storage of rubbish and oily waste.

RULE 5. RESISTANCE BOXES:—

Section *a*. The word "frame" in this section relates to the entire case and surrounding of the rheostat, and not alone to the upholding supports.

RULE 8. MOTORS:—

Section *c*. From the nature of the question, the decision as to what is an *approved* case must be left to the Inspector to determine in each instance.

RULE 9. RESISTANCE BOXES:—

Section *a*. The word "frame" in this section relates to the entire case and surrounding of the rheostat, and not alone to the upholding supports.

RULE 10. OUTSIDE CONDUCTORS:—

Section *a*. Insulation that will be *approved* for service wires must be solid, at least 3-64ths of an inch in thickness, and covered with a substantial braid. It must not readily carry fire, must show an insulating resistance of one megohm per mile after two weeks' submersion in water at 70 degrees Fahrenheit, and three days' submersion in lime water, with a current of 550 volts and after three minutes' electrification. (See List of Wires, page 202).

Section *f*. All joints must be soldered, even if made with the McIntyre or any other patent splicing device. This ruling applies to joints and splices in all classes of wiring covered by these Rules.

RULE 12. INTERIOR CONDUCTORS:—

Section *d*. Insulation that will be *approved* for interior conductors must be solid, at least 3-64th of an inch in thickness, and covered with a substantial braid. It must not readily carry fire, must show an insulating resistance of one megohm per mile after two weeks' submersion in water at 70 degrees Fahrenheit, and three days' submersion in lime water, with a current of 550 volts and after three minutes' electrification. (See List of Wires, page 202).

RULE 13. ARC LAMPS:—

Section *c*. The hand switch to be *approved*, if placed anywhere except on the lamp itself, must comply with requirements for switches on hanger boards as laid down in Section (*g*) of Rule 13.

Section *f*. An *approved* spark arrester is one which will so close the upper orifice of the globe that it will be impossible for any sparks thrown off by the carbons to escape.

RULE 15. OUTSIDE OVERHEAD CONDUCTORS:—

Section *b*. An *approved* fusible cut-out must comply with the sections of Rules 23 and 24 describing fuses and cut-outs.

The cut-out required by this section must be placed so as to protect the switch required by Rule 17.

RULE 16. UNDERGROUND CONDUCTORS:—

Section *b*. The cut-out required by this section must be placed so as to protect the switch.

Rule 17:—

The switch required by this rule to be *approved* must be double pole, must plainly indicate whether the current is "on" or "off," and must comply with Sections a, c, d and e of Rule 26 relating to switches.

Rule 18. Conductors:—

Section a. In so-called "concealed" wiring, molding, and conduit work, *and* in places liable to be exposed to dampness, the insulating covering of the wire, to be *approved*, must be solid, at least 3-64th of an inch in thickness, and covered with a substantial braid. It must not readily carry fire, must show an insulating resistance of one megohm per mile after two weeks' submersion in water at 70 degrees Fahrenheit, and three days' submersion in lime water, with a current of 550 volts and after three minutes' electrification. (See List of Wires, page 202).

For work which is *entirely* exposed to view throughout the whole interior circuits, and not liable to be exposed to dampness, a wire with an insulating covering that will not support combustion, will resist abrasion, is at least 1-16th of an inch in thickness, and thoroughly impregnated with a moisture repellent, will be *approved*.

Section *b*. Second paragraph. Except for floors, *and* for places liable to be exposed to dampness, Glass, Porcelain, *metal-sheathed* Interior Conduit, and Vulca Tube, when made especially for bushings, will be *approved*.

The last two named will not be approved if cut from the usual lengths of tube made for conduit work, nor when made without a head or flange on one end.

Section *f*. All joints must be soldered, even if made with the McIntyre or other patent splicing device. This ruling applies to joints and splices in all classes of wiring covered by these rules.

RULE 21. SPECIAL WIRING:—

Section *b*. The insulating covering of the wire to be *approved* under this section must be solid, at least 3-64th of an inch in thickness, and covered with a substantial braid. It must not readily carry fire, must show an insulating resistance of one megohm per mile after two weeks' submersion in water at 70 degrees Fahrenheit, and three days' submersion in lime water with a current of 550 volts after three minutes' electrification, and must *also* withstand a satisfactory test against such chemical compounds or mixtures as it will be liable to be subjected to in the risk under consideration.

RULE 22. INTERIOR CONDUITS:—

The American Circular Loom Co. Tube, the *brass-sheathed* and the *iron-armored* tubes made by the Interior Conduit and Insulation Company, and the Vulca Tube are approved for the class of work called for in this rule.

NOTE.—The use of two *Standard* wires (see page 202), either separate or twin conductor, in a straight conduit installation is approved in the *iron-armored* conduit of the Interior Conduit and Insulation Co., but not in any of the other approved conduits. (See page 175, Rule 22, *e*.)

RULE 23. DOUBLE POLE SAFETY CUT-OUTS:—

Section *a*. To be *approved*, boxes must be constructed, and cut-outs arranged, whether in a box or not, so as to obviate any danger of the melted fuse metal coming in contact with any substance which might be ignited thereby.

RULE 27. FIXTURE WORK:—

Section *a*. Insulating joints to be *approved* must be entirely made of material that will resist the action of illuminating gases, and will not give way or soften under the heat of an ordinary gas flame. They shall be so arranged that a deposit of moisture will not destroy the insulating effect, and shall have an insulating resistance of 250,000 ohms between the gas pipe attachments, and be sufficiently strong to resist the strain they will be liable to in attachment.

RULE 38. POWER STATIONS:—

Section *a*. Automatic circuit-breakers should be submitted for *approval* before being used.

RULE 44. STORAGE OR PRIMARY BATTERIES:—

Section *b*. Insulators for mounting secondary batteries to be *approved* must be non-combustible, such as glass, or thoroughly vitrified and glazed porcelain.

RULE 45. WIRE PROTECTORS :—

Protectors must have a non-combustible, insulating base, and the cover to be provided with a lock similar to the lock now placed on telephone apparatus or some equally secure fastening, and to be installed under the following requirements:—

1. The Protector to be located at the point where the wires enter the building, either immediately inside or outside of the same. If outside, the Protector to be inclosed in a metallic waterproof case.

2. If the Protector is placed inside of building, the wires of the circuit from the support outside to the binding posts of the Protector to be of such insulation as is approved for service wires of electric light and power, and the holes through

the outer wall to be protected by bushing the same as required for electric light and power service wires.

3. The wire from the point of entrance to the Protector to be run in accordance with rules for high potential wires; i. e., free of contact with building, and supported on non-combustible insulators.

4. The ground wire shall be insulated, not smaller than No. 16 B. & S. gauge. This ground wire shall be kept at least three (3) inches from all conductors, and shall never be secured by uninsulated double-pointed tacks.

5. The ground wire shall be attached to a water pipe if possible; otherwise may be attached to a gas pipe. The ground wire shall be carried to and attached to the pipe outside of the first joint or coupling inside the foundation walls, and the connection shall be made by soldering, if possible. In the absence of other good ground, the ground shall be made by means of a metallic plate or a bunch of

wires buried in a permanently moist earth.

MATERIALS:—

The following are given as a list of NON-COMBUSTIBLE, NON-ABSORPTIVE, INSULATING materials, and are listed here for the benefit of those who might consider hard rubber, fibre, wood, and the like as fulfilling the above requirements. Any other substance, which it is claimed should be accepted, must be forwarded for testing before being put on the market :—

1. Thoroughly vitrified and glazed Porcelain.
2. Glass.
3. Slate without metal veins.
4. Pure Sheet Mica.
5. Marble (filled).
6. Lava (certain kinds of).
7. Alberene Stone.

WIRES:—

The following list of wires have been tested, and found to comply with the requirements for an approved insulation under Rule 10 (*a*), Rule 12 (*d*), and Rule 18 (*a*):

Acme.

Ajax.
Americanite.
Bishop.
Canvasite.
Clark.
Columbia.
Crescent.
Crown.
Edison Machine.
Globe.
Grimshaw (white core).
Habirshaw (red core.)
Kerite.
National India Rubber Co. (N. I. R.).
Okonite.
Paranite.
Raven Core.
Safety Insulated { Requa white core / Safety black core }
Salamander (rubber covered).
Simplex (caoutchouc).
U. S. (General Elec. Co.)

None of the above wires to be used unless protected with a substantial *braided* outer covering.

THE UNIFORM CONTRACT.

FORM OF CONTRACT ADOPTED AND RECOMMENDED FOR GENERAL USE
BY THE
AMERICAN INSTITUTE OF ARCHITECTS
AND THE
NATIONAL ASSOCIATION OF BUILDERS.

THIS AGREEMENT, made the........ day of..........in the year one thousand eight hundred and ninety......by and between............................ party of the first part (hereinafter designated the Contractor), and............... party of the second part (hereinafter designated the Owner),

WITNESSETH that the Contractor, in consideration of the fulfillment of the agreements herein made by the Owner, agrees with the said Owner, as follows:

ARTICLE I. The Contractor under the

direction and to the satisfaction of........
................Architects, acting for the purposes of this contract as agents of the said Owner, shall and will provide all the materials and perform all the work mentioned in the specifications and shown on the drawings prepared by the said Architects for the...................... which drawings and specifications are identified by the signatures of the parties hereto.

Art. ii. The Architects shall furnish to the Contractor such further drawings or explanations as may be necessary to detail and illustrate the work to be done, and the Contractor shall conform to the same as part of this contract so far as they may be consistent with the original drawings and specifications referred to and identified, as provided in Art. i.

It is mutually understood and agreed that all drawings and specifications are and remain the property of the Architects.

Art. iii. No alterations shall be made in the work shown or described by the drawings and specifications, except upon a written order of the Architects, and when so made, the value of the work added or omitted shall be computed by the Architects, and the amount so ascertained shall be added to or deducted from the contract price. In the case of dissent from such award by either party hereto, the valuation of the work added or omitted shall be referred to three (3) disinterested Arbitrators, one to be appointed by each of the parties to this contract, and the third by the two thus chosen; the decision of any two of whom shall be final and binding, and each of the parties hereto shall pay one-half of the expenses of such reference.

ART. IV. The Contractor shall provide sufficient, safe and proper facilities at all times for the inspection of the work by the Architects or their authorized representatives. He shall, within twenty-four hours after receiving written notice from the Architects to that effect, proceed to remove from the grounds or buildings all materials condemned by them, whether worked or unworked, and to take down all portions of the work which the Architects shall by like written notice condemn as unsound or improper, or as in any way failing to conform to the drawings and specifications.

ART. V. Should the Contractor at any time refuse or neglect to supply a sufficiency of properly skilled workmen, or of materials of the proper quality, or fail in any respect to prosecute the work with promptness and diligence, or fail in the performance of any of the agreements herein contained, such refusal, neglect or failure being certified by the Architects, the Owner shall be at liberty, after...... days' written notice to the Contractor, to provide any such labor or materials, and to deduct the cost thereof from any money then due or thereafter to become due to the Contractor under this contract; and if the Architects shall certify that such refusal, neglect or failure is sufficient ground for such action, the Owner shall also be at liberty to terminate the employment of the Contractor for the said work and to enter upon the premises and take possession, for the purpose of completing the work comprehended under this contract, of all materials, tools and appliances thereon, and to employ any other person or persons to finish the work, and to provide the materials therefor; and in case of such discontinuance of the employment of the Contractor he shall not be entitled to receive any further payment under this contract until the said work shall be wholly finished, at which time, if the unpaid balance of the amount to be paid under this contract shall exceed the expense incurred by the Owner in finishing the work, such excess shall be paid by the Owner to the Contractor, but if such expense shall exceed such unpaid balance, the Contractor shall pay the difference to the Owner. The expense incurred by the Owner as herein provided, either for furnishing materials or for finishing the work, and any damage incurred through such default, shall be audited and certified by the Architects, whose certificate thereof shall be conclusive upon the parties.

ART. VI. The Contractor shall complete the several portions, and the whole of the work comprehended in this agreement by and at the time or times hereinafter stated............................
provided that...........................

ART. VII. Should the Contractor be obstructed or delayed in the prosecution or completion of his work by the act, neglect, delay or default of the Owner, or the Architects, or of any other contractor employed by the Owner upon the work, or by any damage which may happen by fire, lightning, earthquake or cyclone, or by the abandonment of the work by the employes through no default of the Contractor, then the time herein fixed for the completion of the work shall be extended for a period equivalent to the time lost by reason of any or all of the causes aforesaid; but no such allowance shall be made unless a claim therefor is presented in writing to the Architects within twenty-four hours of the occurrence of such delay. The duration of such extension shall be certified to by the Architects, but appeal from their decision may be made to arbitration, as provided in Art. III of this contract.

ART. VIII. The Owner agrees to provide all labor and materials not included in this contract in such manner as not to delay the material progress of the work, and in the event of failure so to do, thereby causing loss to the Contractor, agrees that he will reimburse the Contractor for such loss; and the Contractor agrees that if he shall delay the material progress of the work so as to cause any damage for which the Owner shall become liable (as above stated), then he shall make good to the Owner any such damage. The amount of such loss or damage to either party hereto shall, in every case, be fixed and determined by the Architects or by arbitration, as provided in Art. III of this contract.

Art. ix. It is hereby mutually agreed between the parties hereto that the sum to be paid by the Owner to the Contractor for said work and materials shall be $.............., subject to additions and deductions as hereinbefore provided, and that such sum shall be paid in current funds by the Owner to the Contractor in installments, as follows:

....

The final payment shall be made within ....days after this contract is fulfilled.

All payments shall be made upon written certificates of the Architects to the effect that such payments have become due.

If at any time there shall be evidence of any lien or claim for which, if established, the Owner or the said premises might become liable, and which is chargeable to the Contractor, the Owner shall have the right to retain out of any payment then due or thereafter to become due an amount sufficient to completely indemnify him against such lien or claim. Should there prove to be any such claim after all payments are made, the Contractor shall refund to the Owner all moneys that the latter may be compelled to pay in discharging any lien on said premises made obligatory in consequence of the Contractor's default.

ART. X. It is further mutually agreed between the parties hereto that no certificate given or payment made under this contract, except the final certificate or final payment, shall be conclusive evidence of the performance of this contract, either wholly or in part, and that no payment shall be construed to be an acceptance of defective work or improper materials.

ART. XI. The Owner shall during the progress of the work maintain full insurance on said work, in his own name and in the name of the Contractor, against loss or damage by fire. The policies shall cover all work incorporated in the building, and all materials for the same in or about the premises, and shall be made payable to the parties hereto, as their interest may appear..............

ART. XII. The said parties for themselves, their heirs, executors, administrators and assigns, do hereby agree to the full performance of the covenants herein contained.

IN WITNESS WHEREOF, the parties to these presents have hereunto set their hands and seals, the day and year first above written.

In presence of

....
{ L. S. } ....
....
....

COPYRIGHTED 1893.

[NOTE: Persons desiring to use this form of contract (printed here by special permission) can obtain printed copies from the publishers, the Inland Architect Press, 19 Tribune Building, Chicago, who will furnish prices upon application.]

Elementary Electro=Technical Series.

BY

EDWIN J. HOUSTON, Ph.D.,

AND

A. E. KENNELLY, Sc.D.

Alternating Electric Currents.	Electric Incandescent Lighting.
Electric Heating.	Electric Motor.
Electromagnetism.	Electric Street Railways.
Electricity in Electro=Therapeutics.	Electric Telephony.
Electric Arc Lighting.	Electric Telegraphy.

Cloth. Price per volume, $1.00.

The publication of this series of elementary electro-technical treatises on applied electricity has been undertaken to meet a demand which is believed to exist on the part of the public and others for reliable information regarding such matters in electricity as cannot be readily understood by those not specially trained in electro-technics. The general public, students of elementary electricity and the many interested in the subject from a financial or other indirect connection, as well as electricians desiring information in other branches than their own, will find in these works precise and authoritative statements concerning the several branches of applied electrical science of which the separate volumes treat. The reputation of the authors and their recognized abilities as writers, are a sufficient guarantee for the accuracy and reliability of the statements contained. The entire issue, though published in a series of ten volumes, is nevertheless so prepared that each book is complete in itself and can be understood independently of the others. The volumes are profusely illustrated, printed on a superior quality of paper, and handsomely bound in covers of a special design.

Copies of this or any other electrical book published will be sent by mail, POSTAGE PREPAID, *to any address in the world on receipt of price.*

The W. J. Johnston Company, Publishers,
253 BROADWAY, NEW YORK.

THIRD EDITION. GREATLY ENLARGED.

A DICTIONARY OF
Electrical Words, Terms, and Phrases.

By EDWIN J. HOUSTON, Ph.D. (Princeton),

AUTHOR OF

Advanced Primers of Electricity; Electricity One Hundred Years Ago and To-day, etc., etc., etc.

Cloth. *667 large octavo pages, 582 Illustrations.*
Price, $5.00.

Some idea of the scope of this important work and of the immense amount of labor involved in it, may be formed when it is stated that it contains definitions of about 6000 distinct words, terms, or phrases. The dictionary is not a mere word-book; the words, terms, and phrases are invariably followed by a short, concise definition, giving the sense in which they are correctly employed, and a general statement of the principles of electrical science on which the definition is founded. Each of the great classes or divisions of electrical investigation or utilization comes under careful and exhaustive treatment; and while close attention is given to the more settled and hackneyed phraseology of the older branches of work, the newer words and the novel departments they belong to are not less thoroughly handled. Every source of information has been referred to, and while libraries have been ransacked, the note-book of the laboratory and the catalogue of the wareroom have not been forgotten or neglected. So far has the work been carried in respect to the policy of inclusion that the book has been brought down to date by means of an appendix, in which are placed the very newest words, as well as those whose rareness of use had consigned them to obscurity and oblivion. As one feature, an elaborate system of cross-references has been adopted, so that it is as easy to find the definitions as the words, and *aliases* are readily detected and traced. The typography is excellent, being large and bold, and so arranged that each word catches the eye at a glance by standing out in sharp relief from the page.

Copies of this or any other electrical book published will be sent by mail, POSTAGE PREPAID, *to any address in the world, on receipt of price.*

The W. J. Johnston Company, Publishers,
253 BROADWAY, NEW YORK.

Electrical Power Transmission.

By LOUIS BELL, Ph.D.

Uniform in size with "The Electric Railway in Theory and Practice." Price, $2.50.

The plan of the work is similar to that of "The Electric Railway in Theory and Practice," the treatment of the subject being non-mathematical and not involving on the part of the reader a knowledge of the purely scientific theories relating to electrical currents. The book is essentially practical in its character, and while primarily an engineering treatise, is also intended for the information of those interested in electrical transmission of power, financially or in a general way. The author has a practical acquaintance with the problems met with in the electrical transmission of energy from his connection with many of the most important installations yet made in America, and in these pages the subject is developed for the first time with respect to its practical aspects as met with in actual work. The first two chapters review the fundamental principles relating to the generation and distribution of electrical energy, and in the three succeeding ones their methods of application with both continuous and alternating currents are described. The sixth chapter gives a general discussion of the methods of transformation, the various considerations applying to converters and rotary transformers being developed and these apparatus described. In the chapter on prime movers various forms of water-wheels, gas and steam engines are discussed with respect to their applicability to the purpose in view, and in the chapter on hydraulic development the limitations that decide the commercial availability of water power for electrical transmission of power are pointed out in detail. The five succeeding chapters deal with practical design and with construction work—the power-house, line, and centres of distribution being taken up in turn. The chapter on the latter subject will be found of particular value, as it treats for the first time in a thorough and practical manner one of the most difficult points in electrical transmission. The chapter on commercial data contains the first information given as to costs, and will, therefore, be much appreciated by engineers and others in deciding as to the commercial practicability of proposed transmission projects. This is the first work covering the entire ground of the electrical transmission of power that has been written by an engineer of wide practical experience in all of the details included in the subject, and thus forms a valuable and much-needed addition to electrical engineering literature.

Copies of this or any other electrical book published will be sent by mail, POSTAGE PREPAID, to any address in the world, on receipt of price.

The W. J. Johnston Company, Publishers,

253 BROADWAY, NEW YORK.

The Theory and Calculation of
Alternating=Current Phenomena.

BY

CHARLES PROTEUS STEINMETZ.

Cloth. Price, $2.50.

    This is the first work yet written in any language dealing in a complete and logical manner with all of the phenomena of alternating currents and their calculation in the design of alternating-current machinery. In the first six chapters the various primary conceptions and methods of treatment are developed, the use of complex quantities being explained in a remarkably clear and effective manner. The various alternating-current phenomena are then taken up in turn and the more complex parts of the subject approached so gradually and with such a logical preparation that but little if any difficulty will be met in their understanding. The remaining chapters of the book, forming half of its contents, are devoted to the methods of calculation of transformers, simple alternating and polyphase generators and motors, all of the various reactions involved being thoroughly analyzed and discussed. The work contains the very latest knowledge relating to alternating-current phenomena, much of which is original with the author, and here appears for the first time in book form. The high authority of the author on the questions of which he treats, and the original methods which he pursues in their exposition, give this work a character which will assign it to a high place in electrical literature, in which it promises to rank as a classic.

    Copies of this or any other electrical book published will be sent by mail, POSTAGE PREPAID, *to any address in the world, on receipt of price.*

The W. J. Johnston Company, Publishers,
253 BROADWAY, NEW YORK.

Lessons in Electricity and Magnetism.

BY

Prof. ERIC GERARD,

DIRECTOR OF

L'Institut Electrotechnique Montefiore, University of Liege, Belgium.

TRANSLATED UNDER THE DIRECTION OF

LOUIS DUNCAN, Ph.D.,

Johns Hopkins University.

With American Additions as follows: A Chapter on the Rotary Field, by Dr. Louis Duncan; A Chapter on Hysteresis, by Charles Proteus Steinmetz; A Chapter on Impedance, by A. E. Kennelly; A Chapter on Units, by Dr. Cary T. Hutchinson.

Cloth. Price, $2.50.

As a beautifully clear treatise for students on the theory of electricity and magnetism, as well as a résumé for engineers of electrical theories that have a practical bearing, the work of Professor Gerard has been without a rival in any language. As a text-book of reference it has been largely used in American colleges, the logical methods of the author and his faculty of lucid expression and illustration simplifying to students in a remarkable manner the understanding of the various subjects treated. The scope of the present translation has been limited to those parts of the original work treating of theory alone, as the practical portions would not have the same value for American students as for those to whom the book was originally addressed. In order to make it a treatise comprehensive of all electrical theory having a bearing on practical work, and to bring the subject-matter up to date, several chapters written by American authors are added. As will be seen above, the authors of these chapters are authorities on the several subjects with which they deal, and the work as thus extended forms the most complete treatise yet published relating particularly to electrical theory as it enters into the domain of the engineer.

Copies of this or any other electrical book published will be sent by mail, POSTAGE PREPAID, *to any address in the world, on receipt of price.*

The W. J. Johnston Company, Publishers,

253 BROADWAY, NEW YORK.

Publications of The W. J. Johnston Co.

The Electrical World. An Illustrated Weekly Review of Current Progress in Electricity and its Practical Applications. Annual subscription.................... $3.00

Electric Railway Gazette. An Illustrated Weekly Record of Electric Railway Practice and Development. Annual subscription.. 3.00

Johnston's Electrical and Street Railway Directory. Published annually 5.00

The Telegraph in America. By Jas. D. Reid. 894 royal octavo pages, handsomely illustrated. Russia.. 7.00

Dictionary of Electrical Words, Terms and Phrases. By Edwin J. Houston, Ph.D. Third edition. Greatly enlarged. 667 double column octavo pages, 582 illustrations. 5.00

The Electric Motor and Its Applications. By T. C. Martin and Jos. Wetzler. With an appendix on the Development of the Electric Motor since 1888, by Dr. Louis Bell. 315 pages, 353 illustrations 3.00

The Electric Railway in Theory and Practice. The First Systematic Treatise on the Electric Railway. By O. T. Crosby and Dr. Louis Bell. Second edition, revised and enlarged. 416 pages, 183 illustrations 2.50

Alternating Currents. An Analytical and Graphical Treatment for Students and Engineers. By Frederick Bedell, Ph.D., and Albert C. Crehore, Ph.D. Second edition. 325 pages, 112 illustrations 2.50

Practical Calculation of Dynamo-Electric Machines. A Manual for Electrical and Mechanical Engineers, and a Text-book for Students of Electro-technics. By A. E. Wiener 2.50

Gerard's Electricity. With chapters by Dr. Louis Duncan, C. P. Steinmetz, A. E. Kennelly and Dr. Cary T. Hutchinson. Translated under the direction of Dr. Louis Duncan 2.50

Electrodynamic Machinery. By Edwin J. Houston, Ph.D., and A. E. Kennelly, D.Sc. 2.50

The Theory and Calculation of Alternating Current Phenomena. By Charles Proteus Steinmetz 2.50

Central Station Bookkeeping. With Suggested Forms. By H. A. Foster 2.50

Continuous Current Dynamos and Motors. An Elementary Treatise for Students. By Frank P. Cox, B.S. 271 pages, 83 illustrations................................ 2.00

Electricity at the Paris Exposition of 1889. By Carl Hering. 250 pages, 62 illustrations 2.00

Electric Lighting Specifications for the use of Engineers and Architects. Third edition, entirely re-written. By E. A. Merrill.. 1.50

The Quadruplex. By Wm. Maver, Jr., and Minor M. Davis................................ 1.50

The Elements of Static Electricity, with Full Descriptions of the Holtz and Topler Machines. By Philip Atkinson, Ph.D. Second edition. 228 pages, 64 illustrations. 1.50

Lightning Flashes. A Volume of Short, Bright and Crisp Electrical Stories and Sketches. 160 pages, copiously illustrated 1.50

A Practical Treatise on Lightning Protection. By H. W. Spang. 180 pages, 28 illustrations.. 1.50

Electricity and Magnetism. Being a Series of Advanced Primers. By Edwin J. Houston, Ph.D. 306 pages, 116 illustrations..................................... 1.00

Electrical Measurements and Other Advanced Primers of Electricity. By Edwin J. Houston, Ph.D. 429 pages, 169 illustrations.................................. 1.00

The Electrical Transmission of Intelligence and Other Advanced Primers of Electricity. By Edwin J. Houston, Ph.D. 330 pages, 88 illustrations................. 1.00

Electricity One Hundred Years Ago and To-day. By Edwin J. Houston, Ph.D. 179 pages, illustrated... 1.00

Alternating Electric Currents. By E. J. Houston, Ph.D., and A. E. Kennelly, D.Sc. (Electro-Technical Series)	1.00
Electric Heating. By E. J. Houston, Ph.D., and A. E. Kennelly, D.Sc. (Electro-Technical Series)	1.00
Electromagnetism. By E. J. Houston, Ph.D., and A. E. Kennelly, D.Sc. (Electro-Technical Series)	1.00
Electricity in Electro-Therapeutics. By E. J. Houston, Ph.D., and A. E. Kennelly, D.Sc. (Electro-Technical Series)	1.00
Electric Arc Lighting. By E. J. Houston, Ph.D., and A. E. Kennelly, D.Sc. (Electro-Technical Series)	1.00
Electric Incandescent Lighting. By E. J. Houston, Ph.D., and A. E. Kennelly, D.Sc. (Electro-Technical Series)	1.00
Electric Motors. By E. J. Houston, Ph.D., and A. E. Kennelly, D.Sc. (Electro-Technical Series)	1.00
Electric Street Railways. By E. J. Houston, Ph.D., and A. E. Kennelly, D.Sc. (Electro-Technical Series)	1.00
Electric Telephony. By E. J. Houston, Ph.D., and A. E. Kennelly, D.Sc. (Electro-Technical Series)	1.00
Electric Telegraphy. By E. J. Houston, Ph.D., and A. E. Kennelly, D.Sc. (Electro-Technical Series)	1.00
Alternating Currents of Electricity. Their Generation, Measurement, Distribution and Application. Authorized American edition. By Gisbert Kapp. 164 pages, 37 illustrations and two plates	1.00
Electric Railway Motors. By Nelson W. Perry. 256 pages, many illustrations	1.00
Recent Progress in Electric Railways. Being a Summary of Current Advance in Electric Railway Construction, Operation, Systems, Machinery, Appliances, etc. Compiled by Carl Hering. 386 pages, 110 illustrations	1.00
Original Papers on Dynamo Machinery and Allied Subjects. Authorized American edition. By John Hopkinson, F.R.S. 249 pages, 90 illustrations	1.00
Davis' Standard Tables for Electric Wiremen. With Instructions for Wiremen and Linemen, Rules for Safe Wiring, etc. Fourth edition. Revised by W. D. Weaver,	1.00
Universal Wiring Computer, for Determining the Sizes of Wires for Incandescent Electric Lamp Leads, and for Distribution in General Without Calculation. By Carl Hering	1.00
Experiments With Alternating Currents of High Potential and High Frequency. By Nikola Tesla. 146 pages, 30 illustrations	1.00
Lectures on the Electro-Magnet. Authorized American edition. By Prof. Silvanus P. Thompson. 287 pages, 75 illustrations	1.00
Dynamo and Motor Building for Amateurs. With Working Drawings. By Lieutenant C. D. Parkhurst	1.00
Reference Book of Tables and Formulæ for Electric Street Railway Engineers. By E. A. Merrill	1.00
Practical Information for Telephonists. By T. D. Lockwood. 192 pages	1.00
Wheeler's Chart of Wire Gauges	1.00
A Practical Treatise on Lightning Conductors. By H. W. Spang. 48 pages, 10 illustrations	.75
Proceedings of the National Conference of Electricians. 300 pages, 23 illustrations,	.75
Tables of Equivalents of Units of Measurement. By Carl Hering	.50

Copies of any of the above books, or of any other electrical book published, will be sent by mail, POSTAGE PREPAID, *to any address in the world on receipt of price.*

The W. J. Johnston Company, Publishers,

253 BROADWAY, NEW YORK.

THE PIONEER ELECTRICAL JOURNAL OF AMERICA.

READ WHEREVER THE ENGLISH LANGUAGE IS SPOKEN.

THE ELECTRICAL WORLD

Is the largest, most handsomely illustrated, and most widely circulated electrical journal in the world.

It should be read not only by every ambitious electrician anxious to rise in his profession, but by every intelligent American.

It is noted for its ability, enterprise, independence and honesty. For thoroughness, candor and progressive spirit it stands in the foremost rank of special journalism.

Always abreast of the times, its treatment of everything relating to the practical and scientific development of electrical knowledge is comprehensive and authoritative. Among its many features is a weekly *Digest of Current Technical Electrical Literature*, which gives a complete *résumé* of current original contributions to electrical literature appearing in other journals the world over.

Subscription { including postage in the U. S., Canada, or Mexico. } $3 a Year.

May be ordered of any Newsdealer at 10 cents a week.

Cloth Binders for *THE ELECTRICAL WORLD* postpaid, $1.00.

The W. J. Johnston Company, Publishers,
253 BROADWAY, NEW YORK.

www.ingramcontent.com/pod-product-compliance
Lightning Source LLC
Chambersburg PA
CBHW031812230426
43669CB00009B/1115